THE POWER
OF GEMS
AND CRYSTALS

THE POWER OF GEMS AND CRYSTALS

How they can transform your life

SOOZI HOLBECHE

PIATKUS

For Inge and Desmond,
with love

(*Key to front jacket illustration*)
Clockwise from left: Rose Quartz,
Quartz Cluster, Cut Amethyst Ball,
Amethyst, Smokey Quartz, Labradorite,
Agate, Snowflake Obsidian, (Central
Stone — Diopside).

First published in Great Britain in 1989
by Judy Piatkus (Publishers) Limited,
5 Windmill Street, London W1P 1HF

British Library Cataloguing in Publication Data

Holbeche, Soozi
    The power of crystals and gems
    1. Crystals
    I. Title
    548

ISBN 0−86188−859−6

Edited by Martin Noble
Design and illustrations by Zena Flax
Inside colour photographs by Mark Van Aardt
Cover photographs by Tim Imrie
Typeset by Action Typesetting Limited, Gloucester
Printed and bound in Great Britain by
Mackays of Chatham plc

Contents

With love and appreciation to Lynn Buess and all my fellow crystal workers who over the years have taught, inspired and guided me. And a special thank you to Desmond for letting me bounce the book off him.

I dedicate this book to crystals, for all they embody of the spirit of love and light, and for all they have done for me:

The disciple must have a heart as pure as a crystal
A mind as bright as the sun
A soul as vast as the universe
And a spirit as powerful as God and one with God

Peter Deunov

Legend has it that Hercules dropped the crystal of truth over Mount Olympus, where it shattered into a million irridescent pieces. Whenever we handle crystals today, we are picking up glittering fragments of that same eternal truth.

My life with crystals and stones

My life began on a tea-plantation in Sri Lanka, or Ceylon as it used to be known. This jewel-like island lies off the tip of India. King Solomon is said to have sent some of his subjects to it for precious stones with which to woo the Queen of Sheba. This is the island from which Dutch and Portuguese sailors used to return carrying pouches full of gems, rubies, sapphires, garnets and moonstones.

My earliest memories have a dreamlike quality. I recall the misty blue hills around Kandy, the temples and shrines at the sides of the road alive with flowers, the beautiful colours and graceful movements of the sari-clad, brown-skinned women picking tea. Vivid, too, is the image of the daily procession of elephants being taken to the river to bathe. One of them would stamp its foot like an irate child if the expected lump of sugar wasn't produced promptly at 4 o'clock.

Stones for me, then, were something I fell over when running away from my ayah or picked up to throw when frightened by the high-pitched shouts of the garden-boys finding snakes. Stones were things that grazed my knees, caused me pain and got me into trouble by dirtying my clothes.

One day when running away from my ayah out of the garden and into the street I was caught up in a procession of clapping, chanting people led by saffron-robed monks. I was carried along in the crowd until I found myself pressed against a caged statue of Buddha.

Gazing through the bars of the cage I felt pierced by laser-like beams of light streaming from the Buddha's eyes. It was as if I saw God. For me the statue was alive and something in me awoke. From this moment on I perceived the world in a different way.

I could see the life force or inner essence of everything around me. Trees, rocks, stones, mountains, even the walls of the house, pulsated, breathed, moved. Flowers and trees spoke to me. Thoughts and words flew out of the people around me as streams of colour and energy. I could see what they thought. It did not always match what they said. It was painful and confusing. It seemed as if everyone was telling lies.

I saw past, present and future happening simultaneously. I dreamed that people died, and they often did. I saw through things. So called 'solid' matter such as tables, chairs, trees, houses disintegrated or dissolved into vibrating atoms in front of my eyes. It was as if the molecular structure of the universe, in constant flowing movement of expansion and contraction, opened itself to me.

I would say things and they would happen. I thought I was making things happen – not understanding that it was possible to know things before they happened. I knew nothing of clairvoyance, telepathy, psychic sensitivity. I would 'see souls' in and out of the body – sometimes as lines of energy, sometimes as blue circles, sometimes as shadow outlines of human forms through whom I could see and with whom I could communicate.

If I had questions I was always answered, either by the universe around me or through dreams. Much of what I have learned has come through dreams. Answers to such questions as: What happens when we die? Where does lack of self-worth come from? Why do we come into a physical body? Do we choose our parents?

My dreams and my waking experiences showed me that we write our own script. We direct it, produce it and act in it. We forget that, if the play of our life turns out not to be a box-office success, we can step back and recreate it – by a change of attitude if nothing else.

I felt as if everything wanted to tell me its story. I would see the 'beings' of rocks, mountains and stones. Figures, faces, shapes would emerge and speak. I learned that if I wanted to sit in a particular place I would have to ask the stones in the area for permission. If they said 'no' I would have to start again, moving from rock to rock until one said 'yes'. In the same way I was taught to 'ask' if a stone wanted to come home with me, and not just to take it.

I always loved, but hated picking, flowers. One day when I was sitting in the middle of a flowerbed they suddenly told me that they liked being picked. They said, 'It's just as if you painted a picture and someone liked it so much that they wanted to take it home. You'd be happy, pleased. It does not hurt us any more to be picked than it hurts you to have your hair cut.'

I remember reading that George Bernard Shaw, when asked why he didn't like picking flowers, said 'I like children but I don't chop their heads off.' I still feel like him. I love flowers and do pick them sometimes but I always ask and say thank you afterwards.

It is the same with fruit and vegetables — or even cutting and pruning. Tell the plant what you are going to do and why you need to do it two to three days beforehand. It will then co-operate with you by withdrawing its life-force from the part that is to be removed.

The trees, too, used to call to me. I found hugging trees to be wonderfully comforting and revitalizing. I still do it today. Years ago in Africa, in the beautiful garden of friends at Kloof in Natal, the trees called me outside into a pitch-black night, and made me promise to speak for them whenever I could. They told me that the energy being 'fed' into the planet at this time was for all matter, all life to make a quantum leap in consciousness. They said too, however, that due to pollution — cutting down trees that help provide us with oxygen to breathe, the excessive use of gasoline, the general rape of the planet — the 'ring-pass-not' (whose physical equivalent is the ozone layer) was disintegrating for the first time in man's history. The energy meant for the quantum leap was being used instead to prevent an even faster disintegration. The trees showed me this many years before the hole in the ozone layer was discovered over Antarctica.

I found that if a plant or an animal was ailing I would instinctively put my hands around it and feel a great flow of warmth moving down my arms into my hands and into the plant. My hands tingled.

3

If a friend had a headache or a pain, I instinctively tried to take it away. A ray of light, like a sword, sometimes deep blue, sometimes colourless, streamed from the centre of my forehead and I saw through to the part that was hurting. Occasionally people would say, 'Stop! It's too strong.'

If I ran my hands over an area of pain it would often disappear. If I thought very strongly about someone who was ill, I would find my consciousness inside the part of the body that was suffering.

I discovered that I could think about a plant and 'mentally water' it — in other words, visualize myself watering it — and that this had a restorative effect on it. If I worried about plants not being watered because I was away longer than planned, I have sometimes, while mentally watering and talking to them, found myself inside the plant looking out of the pot — a slightly uncomfortable position when the plant was on a shelf high up!

At no time did I try to do this. The shift of focus just happened. Everything I did was an automatic response.

I learned instinctively that all matter has a kind of life-force that is 'responsive' to other energies, including human ones. All matter can be impregnated or literally changed by the energy of our thoughts, words and actions — whether it be a chair, a rock, a car-engine. I saw atoms and molecules spinning at different vibratory rates, creating matter of greater or lesser density.

According to Fritjof Capra, the acclaimed physicist and author of *The Tao of Physics*:

> we know that all matter is a whirling mass of movement containing no isolated lumps. It is continuous dancing vibratory motion whose rhythmic patterns are determined by molecular atomic and nuclear structures. The atom itself is no more than an area of space wherein electrical forces establish the nature of the nucleus and its associated electrons. Each atom is a field of energy possessing positive, negative and neutral charges capable of producing electric and magnetic forces it is impossible to describe.

This shows how the new physics connects with the teachings of mystics throughout the ages from all traditions.

When my focus shifted I had no control over my 'seeing'. It was as if a switch suddenly went on or off in my brain and I was able to 'tune in' to these different levels of vibration exactly as we tune in to different channels on radio or TV. It was confusing and I became afraid.

There were other insecurities. The upheavals of war which flung me away from home in Sri Lanka to England, the later divorce of my parents, my convent education, drove me very much in upon myself. I lived a double life. A secret inner life and a somewhat traumatic outer life.

My convent school had a particularly strong influence on me. I was one of two non-Catholic girls in a Catholic school. I was seen as a sinner. The nuns told me regularly at morning prayers that my mother was a sinner because she was divorced and that being the daughter of a divorced woman made me a sinner too. I believed it. I felt different. I felt guilty.

I learned to say nothing about what was really going on in my mind. I tried to shut off my psychic sensitivity. However, the more I turned my back on this inner-seeing eye the more difficult my life became.

Externally I developed a shell of brash confidence that caused me to jump into a variety of jobs and situations that seemed glamorous and successful. Fashion, advertising, modelling and marriage – I tried them all. I became chronically accident-prone. If there was a plank to fall off, a flight of steps to trip down, a cupboard to knock myself out on, I did it.

I was also desperately unhappy. My only way of dealing with my problems was to retreat into illness. I did this quite unconsciously. I spent a considerable amount of time in and out of hospital. Only today do I realize that physical illness is a manifestation of an imbalance on another level – be it mental, emotional or spiritual.

I had reached a point of despair where I felt that even if I never set foot outside the front door the ceiling would fall on my head or a ten-ton truck would crash through the kitchen window.

Two events then occurred which totally changed my life. The first was as the result of a series of illnesses and operations.

I had what is now known as an NDE or Near Death Experience. Despite my belief in life after death, I had always had a great fear of dying. Not a fear of being dead but of the act of dying. I imagined that no matter how one died — accident, illness, violence or suicide — there would always be a struggle, a gasping for breath, an unpleasant choking.

However when I actually had an NDE there was absolutely no sense of struggle. I simply found myself out of my body, being drawn through a circle of radiant white light into an unearthly, deep, rich, blue colour — delphinium blue — which lengthened into a tunnel-like space. I had a sensation of floating and of absolute peace and tranquillity. I remember thinking that if only people knew what it was like to die they would not be afraid. I reached a point at which a voice asked me if I thought I had completed what I had come to do. Was I going to leave my son, then aged three, behind? I felt no threat or coercion, no judgement. The voice seemed absolutely to accept that whatever I did was all right but it pointed out that the moment of choice was now. I found this clarification infinitely freeing. Despite a longing to let myself drift into the blue and the light and the love that surrounded me I found myself being drawn backwards into my body.

During the entire episode I was aware of my body as a coat I had shrugged off and left at a dressmaker. What was being done to it was witnessed by me with a detached curiosity as if it were a garment being mended. In speaking to others with similar experiences I find that no matter what is happening to the body during the life versus death trauma one's essence is unaffected. At no time did I choke or gasp, no matter how my body appeared to behave. This experience totally took away my fear of dying. The reminder that I am not my body, that the essential 'me' is something else, freed me to live my life in a different way.

However, I was still very accident-prone. In spite of throwing myself into everyday life I was still dealing with what was rather like an on/off switch in my brain that could suddenly catapult me into a totally different perception of reality. I was still uneasy about it, although beginning to find out, through books and other people with similar powers, that it was not so peculiar. It is an inner, instinctive knowledge or wisdom that most primitive people still use today. It could be termed the inner counterpart of the five external senses. With the emphasis on Aristotelean, left-brain logical thinking, as opposed to right-brain intuition, these senses began to atrophy in Western man centuries ago.

We need to remember that these inner senses are as much a part of the human being as walking and talking, thinking or listening. We all have access to them. Using our imagination, creatively expressing ourselves through colour, movement, words, music, sharing our dreams, and meditation, are all keys to getting in touch with them.

The second major, and initially destructive, event that occurred in my life was an accident in which I smashed my face into the corner of a steel-capped trunk. I was running to answer the 'phone in the house when I tripped and fell.

It was some hours before help came. I crouched in the corner where I had fallen holding my face together. In hospital I went into shock and I did not immediately realize what had happened to me.

The first time I saw myself was about a week later. I had been taken by a nurse to the bathroom. Another patient called, the nurse left me holding the handle of the bathroom door and I turned towards a mirror that ran the length of the wall. I saw a totally inhuman, grotesque creature. A 'blob' without features, the face swollen into a Mickey Mouse-like protuberance.

My stomach rose into my mouth and I thought I was going to vomit. Then enormous pity for the creature I thought was standing behind me flooded me. Gagging, I held my throat and managed to swallow the bile. I turned to see who it was. There was no one there. I did this three times before I stumbled towards the mirror. I raised my hand to my own face and realized that the blob was me.

I then collapsed in despair. I wanted to die. I retreated into my bed and refused to see anyone. I was angry with God, with life and with all those whose lives continued happily around me.

Questions arose in my head. 'Why me? What have I done to deserve this?' Finally this impotent, fist-shaking rage moved me into a defiant, 'I'll-show-you' type of anger. I felt cut off and alone but then I remembered the fire-walkers in Sri Lanka who trod thigh deep in flaming coals. They did not sweat, feel pain or even burn their skin. These were men who had so mastered their reactions that they were able to walk through the flames with no apparent fear or damage.

I decided that if a so-called uneducated, simple man could train his mind so that he could walk through fire then it also ought to be possible for me to train my mind to control my physical reactions.

I realized that as long as I said, 'This has ruined my life', and hid under the bedclothes, I not only divorced myself from what had happened but also gave power to it. When I was finally able to act rather than react I robbed the disaster of its power over me. This discovery helped catalyse a positive change in my attitude to life.

I fought back. I used visualization, affirmation and prayer. In the beginning my affirmations were angry statements of 'I will *not* live in a body looking like this.' My prayers berated rather than beseeched the gods.

I taught myself to meditate. It was not easy. However it put me in touch with the part of my mind that was able to control and use the awareness I had previously feared.

I was told through dreams and meditation, and by the mineral kingdom itself, that it was time to develop a relationship with crystals. I was shown that crystals could help my skin to heal, my scars to knit and that by holding crystals in a certain way I could draw pain away from my face just as I had been able to before, with my hands, when treating people.

I learned that by keeping crystals within a 2–3 ft radius of me, in my purse, under my pillow, in my pocket, their vibration affected the cells of my body. Much later I discovered that scientifically crystals affect our bodies due to their composition. They consist of silicon dioxide and water. Our bodies too contain 70–78 per cent silicon and water.

Crystals are of a very high vibrational frequency (see p.4) which can be measured scientifically. They bring up to their own vibratory level anything within a certain radius. Hence their effect on our body cells. I drank water in which I had soaked a crystal for twenty-four hours and it improved my skin.

My face began to get better. It did not happen overnight. It was hard work. I went through many ups and downs but I finally had what I believe to have been a miraculous healing.

Surrounding myself with crystals not only helped my body but also my mind to be more awake, more open. The crystals did not do it all on their own but I believe that their energy was vital to my cure. They also helped me to see more clearly what I could do for myself by my own will-power.

I realized that if the power of my mind, aligned with the higher mind of God, could literally heal and change the cells of my body then nothing in life was impossible.

I saw that if this miraculous change could happen for me then so could it happen for everyone else. The healing may be more subtle than total physical transformation. It may cause a change of attitude or even release from the physical body through death. After all, ultimate healing is eventual release from the body. We never came to this planet to be permanent residents. We came as visitors to play, to explore and to learn, before going home. Some of us make shorter visits than others and we need to remember that there are other planes of existence and that we shall always see those we love again.

Most of us have no conception of how little we use the wide array of capabilities open to us. I remember a man sitting in an Indian market in a village without water or electricity. In front of him was an 'empty' fish-tank which on closer examination proved to be full of fleas. When asked why they did not jump out he said, 'I have trained them not to.' By putting glass over the tank the fleas apparently jumped and knocked themselves out. When they finally learned to limit their jump he removed the glass lid. Metaphorically speaking, the 'glass lid' of our limited understanding is now being taken off for all humanity. We are now making a quantum evolutionary leap as great as if we were stepping from water on to land and standing for the first time. The glass lid is the limited perception of our capabilities.

Most of us do not even use 10 per cent of our brain. I believe that the average person uses only 5–6 per cent. By thinking differently, by leaping out of the metaphorical fish-tank, we can jump as high as the stars. We can tap into limitless creativity. Creativity that enables us to say, 'I do not like the script I wrote for myself. I am going to do something about it.'

The word disaster literally means (in latin) 'separation from the stars' and it is obvious that much of life on earth today is disastrous – competition, murder, war and pain – separation from light and joy, from love and laughter. We are separate from our own perfection, our own divinity.

Quartz crystals can help us make that quantum leap to the stars. They are like crystallized light – starlight. They are the most evolved stone of the mineral kingdom. The word crystal comes from the Greek *crystallos*, meaning 'frozen water' or 'clear ice'. In ancient times they believed that holy water was poured from the sky by the gods who froze it to preserve it.

In Greek legend, Hercules is supposed to have dropped the crystal of truth while climbing Mount Olympus. It shattered into millions of fragments that were spilled all over the world.

To dream of crystals is to dream of spiritual perfection and truth. In every culture and tradition crystals are revered as being a symbol of divine perfection. Nikola Tesla* wrote 'In a crystal we have the clear evidence of the existence of a formative life principle and though we cannot understand the life of a crystal it is nonetheless a living being.' Like Tesla I was increasingly fascinated by stones, gems and particularly quartz crystals.

After my recovery from my accident I returned to work with theatrical make-up in theatre and television. But there was a continuity in my discoveries about healing. Part of my job involved helping accident victims with severe facial disfigurement so that they could go back into the world. My own experience, though minimal in comparison to the traumas of many of those with whom I worked, enabled me to give this help.

To get more experience with the sculpture of the face I did a lot of voluntary hospital visiting, making-up patients, practising on them. This led me to talk about my discoveries with healing, how I had worked with my face and recovered from NDE. I began working in greater depth with people in all

---

* Nikola Tesla (1856–1943), a Croatian electrical engineer, was a scientific genius. He was able to write complex mathematical equations on an imaginary blackboard in his head. Among his inventions was the first alternating current generator.

sorts of crises, from terminal illness to nervous breakdown.

I found that putting crystals around a dying patient created an atmosphere that was lighter and brighter. If I felt afraid or inadequate I kept a crystal in my right-hand pocket. It helped me to be more assertive. If I wanted to meditate or still my mind I kept the crystal on my left side. Much later I found out that unless one is left-handed, the right side of the brain controls the left side of the body whereas the left side of the brain controls the right side. The right side is the intuitive receptive part of the brain whereas the left side is the rational, logical, expressive part. When I spoke about this in a workshop it prompted a girl to borrow a crystal from me for the evening. When she returned it on the following day to my dismay she had a black eye, however, she was delighted to tell me that, with the crystal held firmly in her right hand she had had the confidence for the first time in ten years of marriage to a bullying husband, to stand up for herself and say, 'You will *never* treat me like this again. If you do I will leave you.'

I discovered that if I held a crystal over a wound and rotated it in an anti-clockwise movement it helped to draw pain out. I sometimes saw light pouring out of the crystal into the wound.

A friend who did this with her dog after accidentally running him over in her car found that his leg, which she had broken, healed much quicker than usual. The dog loved being treated with the crystal. He lay down on his back with his paws in the air when he saw it. Another dog I know used to wear a crystal around his neck, attached to his collar. His owner felt he strayed less. A homeopath found that after a week of giving her animals crystal-treated water they refused to drink the non-crystallized. She felt that they were healthier, more sparky as a result.

I used to have a parrot called P.K. He was brought to me unexpectedly by people I did not know and who obviously wanted to get rid of him in a hurry. I have always loved birds, parrots in particular. P.K. was the fourth to come into my life. Naïvely, I welcomed him without thinking about the speed with which the owners had deposited him and departed. I soon understood why. I was scratching his neck. In apparent ecstasy he suddenly bit me right through to the bone of my right thumb. In agony I lashed out and hit him — for which I was deeply ashamed afterwards. An ornithologist told me that parrots have very long memories and that I would never make friends with this bird now. With tears rolling down my cheeks I apologized to P.K. but reminded him that it was not nice to bite the hand that fed him. I then tried to improve our relationship.

I surrounded his cage with crystals. I meditated with him. I talked to him. I carried him from room to room. I included him in my sessions. I put a crystal in his water and another inside his cage. Things were getting better when the most terrible accident happened. While putting him to bed in his cage one night he caught fire from a candle flame. For a moment too shocked to move I watched flames envelop him. By the time I'd grabbed him and extinguished the fire his feathers were black and charred from top to toe. I thought we would both die of shock.

I took him to bed with me and nursed him all night, keeping a circle of crystals around us both. I gave him drops of crystal water. In the morning he marched into his cage. He spent the entire day pulling the burned feathers out. His skin had not been burned at all. The feathers subsequently grew back as thick and colourful as before. He became my most loving companion. He sat on my shoulders or sat on my head. He wandered around the garden by day and returned to his cage at night. He travelled with me. He eyed my crystals with a beady eye but he tolerated everything that was done to him by them.

I learned that by giving a person who was agitated a crystal massage — by rotating the crystal in a clockwise movement all over the body from a distance of about six inches — a sense of peace and relaxation that often led to sleep was created. I found too that if I was tired and sat with a crystal in my lap for five minutes I was recharged as if I'd plugged into an electric socket.

The awareness that had seemed like a curse earlier in my life became a tool for helping. Once I stopped resisting it I ceased to be accident-prone. My health improved. I saw that the situations in my life had been part of a training. I dropped all other work.

Through touching people's faces I found that my hands instinctively wanted to move to other parts of the body and the kind of 'session' I do today began to evolve. My hands were drawn to areas of blocked energy. As I touched them it was like reading braille. I found memory and emotion locked into specific points of the body that showed the root cause of current problems with health, relationships, work or phobias. Sometimes this involved regression into past lives. As people rediscover these memories and heal them they are able to live their lives no longer victimized — consciously or unconsciously — by the past.

I always surround my patient with crystals because their energy is conducive to healing and clarity of thought. Because have so many in my home people will often step through the door and feel their necks click into place or their spines move before I have even touched them.

I do not believe that healing energy comes from me but through me. I can act as a catalyst that can trigger a person's own capacity to kick into gear — like a car needing a jump start when the battery is flat. In the same way crystals can help kick-start the body to balance itself to a point of optimum efficiency and health. They are transformers of energy and balance everything with which they come into contact.

Crystals are as effective with earth and plants as they are with people and animals. Each one has a unique quality and an individual personality. Just as the plant kingdom responds to acknowledgement from us by giving more flowers, fruit and fragrance, so do crystals flower by developing rainbows as colour prisms. If I am feeling off-colour the colours in my crystals will disappear and because they are attuned to my emotions they will only come back when I myself am feeling better.

My mother has often thought that much of what I do is crazy. I went to visit her one day after she had been away from home for a few weeks. She was upset because the friend who had agreed to water her plants had not done so. The plants were now three-quarters dead. Some were drooping all over the floor. I had with me about forty hand-sized crystals so I gathered the plants together and put the crystals around them. My mother was astonished to see that within thirty minutes the plants had not only responded but were almost recovered. Within an hour they looked as if nothing had ever happened to them. I went upstairs to unpack. When I came down I was in turn amazed to find my mother seated in a chair with her feet on a row of crystals. Obviously they helped her. Since then she keeps one with her wherever she goes and sleeps with it under her pillow.

Crystal power

For many thousands of years rocks, stones and gems have been an endless source of awe and wonder for mankind. Crystal lenses have been found in ruins dating from 3800 BC and quartz is used today for the production of lenses for special optical instruments. Australian and New Guinean tribesmen have used crystals for making rain and the American Indians considered crystals to be the brain cells of Mother Earth.

Ayers Rock, or Uluru, in the centre of Australia, emits the frequency of a giant crystal. The Aboriginals believe it to be a sacred living being which marks the centre of the world. For me the effects of being near Uluru were so powerful that I felt

A crystal cluster

reconnected to ancient echoes of myself. Every cell in my body came alive. My mind opened to the sun, the moon and the stars, my body to the deepest recesses of the earth. This is exactly what quartz crystal does. It expands us, renews us, connects us to all our parts.

Marcel Vogel, who was a research scientist with IBM for twenty-seven years, and responsible for developments such as magnetic coding for computer tapes and phosphors used in colour TV images (phosphors are crystals that have impurities and emit light when excited by heat), discovered, while watching liquid crystals, that if he projected a thought into a crystal before it became solid it took the shape of his thought. For example, if he projected the thought of a tree, the crystal would grow into this shape. This prompted him, as both a material scientist and spiritual investigator, to devote the rest of his life to understanding, measuring and coming to grips with the power of crystals. In an interview in the January/ February 1985 issue of Yoga Journal he said that in all radio transmitting stations a crystal is used as the primary means of communicating. A slab of quartz (silicon dioxide) is cut to a particular dimension and ground down until it emits the desired frequency. If you want 650 kilocycles you grind the crystal down until you get 650 kilocycles.

According to Vogel, the patterns of thought vibration oscillate like a magnetic field. Thought is contained in geometric pattern forms in space, causing that space to oscillate. That oscillation in turn acts on matter that is in space, namely the atomic forms of oxygen, nitrogen and the water molecules, causing them to vibrate and to propagate a series of patterns that move outward into space around the body. This is much the same as the circular pattern formed when a rock is thrown into the water. When we 'think', we generate a pattern and this pattern then oscillates and radiates a field that acts on matter. This is 'the energy that follows thought'.

Quartz watches and clocks get their name from the quartz crystal which is an essential item in their manufacture and which is responsible for the accuracy of their time-keeping.

Frank Dorland, now an established authority on crystals — especially in their cutting and shaping — used to be an art conservator, specializing in religious art. He was then lent the famous Mitchell-Hedges crystal skull (a carved Mayan skull, a similar example of which can be seen in the British Museum). He says, 'if we put our hands close to the skull we could feel a tingling like an electrical current. We saw shapes and shadows moving inside it; we heard voices and music, sometimes there was the fragrance of apple-blossom.' He realized that the skull was carved to perform a series of optical illusions through the use of light. His fascination with, and desire to discover more about, this strange phenomenon led him and his wife to abandon art conservation and specialize in bio-crystallography (the study of the structure of crystals).

Edgar Cayce, well-known psychic and founder of the ARE, (Association for Research and Enlightenment) in America said, in a reading that is recorded in the book *Gems and Stones*,

> We find that the crystal as a stone, or any white
> stone, has a helpful influence — if carried about
> the body; not as an omen, not merely as a 'good
> luck piece' or 'good luck charm' but these
> vibrations that are needed as helpful influences
> for the entity are well to be kept close about the
> body.

In another reading he says,

> As to stones — have near to self, wear preferably
> upon the body, about the neck, the lapis lazuli;
> this preferably encased in crystal. It will be not
> merely as an ornament but as strength from the
> emanation which will be gained by the body
> always from same. For the stone is itself an
> emanation of vibrations of the elements that give
> vitality, virility, strength, and that of assurance in
> self.

In 1980 Dael Walker, Director of the Crystal Awareness Institute in America, rented a booth at the New York State fair and tested people's ability to heal themselves with crystals. Out of the 234 people who were tested for reduction of pain or stiffness, by holding a crystal, 227 confirmed a significant reduction of pain. In an article written for the *Crystal Source Book* in 1987, he said,

> the crystal responds to thoughts and emotions and
> interacts with the mind. It increases thought
> energy and emotional power. Reduction of stress
> and pain and accelerated healing are ordinary
> parts of crystal energy balancing. We have whole
> systems of simple but extremely effective methods
> to reduce healing time by at least half of the
> accepted standards.

There is a growing awareness of the power of crystals by the media. To give a few examples: an article in *Time* Magazine, 19 January 1987, entitled 'Rock Power for Health and Healing', talks about a businessman putting a crystal in his pocket to enhance concentration; a woman treating her bronchitis with an amethyst; the benefits of drinking gem water; and interior decorators putting crystals in homes and offices. Another article, 'Path to Power is Crystal Clear' (*USA Today*, 27 January 1987), discussed the fact that crystals weighing between 200 and 4,000 lbs were becoming an increasingly commonplace part of office equipment.

Shirley MacLaine is a convinced user of crystals. She was shown on the front cover of *Time's* 7 December 1987 issue holding a crystal cluster to illustrate a seven-page article inside the magazine. Tina Turner, the world famous rock-star, is another crystal fan who never travels without crystals. The British actor Charles Dance, and his wife Jo, also use crystals.

During a conference of dowsers at the University of California in Santa Cruz in 1985 tests were done on the effects of drinking ordinary and crystal-charged water. It was found that within 45 seconds of drinking the crystal-charged water there was a dramatic expansion — to 15 feet or more — of the electro-magnetic energy field around the body. Drinking normal water showed no change.

Crystals turn on a light in our lives that makes everything clear. It is as if one sits in a room day after day thinking it is tidy and dust-free. Suddenly someone comes in with a giant flash-light. As the beam illuminates the room we suddenly see the cobwebs on the ceiling, the marks on the floor, the dust on the table leg. They were there all the time; we just failed to see them.

Crystals do not make things happen as some people believe. They bring clarity. In addition they stimulate our inner resources to cope. From being nervy, accident-prone, emotionally super-sensitive, I now live my life in a state of peace and balance. Since I discovered crystals I sleep better, work better. I am in harmony with myself and all that goes on around me. I no longer do six things at once. I accomplish what I set out to do.

Crystals and stones: a history

Since the beginning of time, mountains and stones, rocks and gems have played a vital role in man's evolution. Precious stones, whether discovered in the tombs of Egyptian pharaohs, the ruins of Incan temples, or merely dug out of the ground, have exerted an endless fascination.

The layered crust of the earth contains fossilized plants and animals, minerals that illustrate nature's way of combining elements into compounds and crystal structures, human bones and architecture, even meteorites that have fallen from the planets. Scientists studying this layering of rock can find information about not only the earth but the universe too.

History is rich with stories about the mystical and magical power of stones. Fables about the lost civilization of Atlantis suggest that crystal power was used in much the same way that we use electricity today. The Papyrus Ebers (an Ancient Egyptian work of 1500 BC, discovered during the excavation of a pyramid) detailed the curative use of gems and minerals while the oldest formulas preserved for us came from the Sumerians who preceded the Babylonian culture.

The Babylonians wrote of trees on which grew precious stones, and the Hindu Puranas describe Krishna living in a city, 'furnished with cupolas of rubies and diamonds − with emerald pillars and courtyards of rubies. It had crossroads decked with sapphires and highways blazing with gems.' Amber was written about in Homer's Odyssey while Pliny and Paracelsus both described the healing power of stones.

Even prehistoric man was interested in rocks and minerals.

He searched for the stones that would make the best tools and weapons. He ground them up to make colour for painting himself or his pictures. He wore them as personal adornment. As people moved from a nomadic existence to a more settled social order, minerals and precious stones began to be discovered. Gradually decoration for every part of the body developed, including collars, anklets, armlets and headbands.

Many ancient legends and stories trace rocks and stones back to the Earth's beginnings. One relates that the stone foundation is the rock on which the universe is based. It is the keystone of the Earth and the source of the waters of life — the rock which prevails against Hades and the powers of the underworld.

Paul Solomon, the American mystic teacher and healer, wrote that in the beginning the world was created for us to enjoy. It was a garden in which we could play and explore. He says that the sons and daughters of God who first came to Earth were not in a physical body. He describes these etheric forms of life as being able to project themselves into a rock, a stone or a tree in order to experience what it felt like to be one.

Maybe this is what led to the belief in many primitive cultures that stones could give birth to people and that people could give birth to stones. Stones and rocks are believed by most ancient peoples to be inhabited by spirits and are therefore considered to be sacred.

A Cherokee Indian told a story about a member of his tribe who owned a crystal: he fed it by rubbing it all over with the blood of a deer. It was considered to be a living entity that required nourishment. Ancient history shows that Mexicans thought of blood as being the water of precious stones.

Many cultures believed that stones and rocks had a life-giving potency which would provide the wearer with strength if worn or carried. Over the years this led to the use of gems and stones for healing various conditions in the body until it became a highly developed science.

The Papyrus Ebers recommends treatment such as sapphires for diseases of the eye, emeralds as a laxative and for dysentery, rubies as a cure for spleen and liver problems, while the amethyst was to be used as an antidote for snakebite — 'when used in a pendant suspended on a dog-hair cord'.

When Pope Clement died in 1534 he was said to have taken as medicine pulverized gems valued at 40,000 ducats. In the seventeenth century writer William Salmon gave a recipe to improve the skin: 'Dissolve pearls in juice of lemons, or distilled vinegar digested in horse dung which will send forth the clear oil which will swim to top.' It is not clear if this mixture was to be applied or swallowed!

In 1609 Anselmus de Boot, the Court Physician of Rudolph II of Germany, wrote about good and bad angels who, 'by the special grace of God and for the preservation of man were able to enter precious stones — to guard them from danger'. The influence exerted by precious stones was accepted without question in medieval times.

Stones have played a key part in the development of various world religions. Jesus said to Peter (Peter meaning rock) that he was the rock on which Christ would build his church. Islam is founded on what happened to Mohammed when the stones spoke to him. Mohammed was coming back to his village from the desert when he heard voices announcing to him that he was to receive the Qur'an from Allah. He turned around and could see no one. The message came from the stones. One of the five pillars of Islam says that every Muslim must try to visit Mecca at least once in his life on a Hadj — a pilgrimage. The central focus of his journey is to visit the Ka'aba, a huge black stone which is the point of communion between man and Allah.

In 331 BC a Baelytic stone — a stone that denotes a place where divinity dwells and that can speak and prophesy — changed the course of history. Alexander the Great, after he defeated Darius went to ask the Oracle of Ammon at the oasis of Siwa in Egypt to confirm that he was the son of Zeus. Alexander never told what the Oracle revealed to him but he went on to conquer the whole known world. The conquests of Alexander also introduced many gems to Greece.

In both Ancient Egypt and Greece priestesses were trained to become oracles and frequently spent the rest of their lives incarcerated within a small stone structure. The only opening was a narrow slit through which they were fed. This added to the belief that stones contained 'an indwelling spirit'.

Moses received the Ten Commandments on tablets of stone

The High Priest's breastplate (artist's impression)

and some writings suggest that these stones were sapphires. The Jewish historian, Josephus, in describing the jewel-encrusted breastplate of the High Priest, who lived at the time of Moses, said, 'there emanated a light from the stones as often as God was present'.

The use of stones to decorate the dead can be traced back to remote time. The Egyptians engraved verses from the Book of the Dead on gems such as carnelian, green jasper and lapis lazuli. The form of an eye made from lapis lazuli was considered particularly powerful as it symbolized Ra, the Sun God. Major civilizations associated gems with the deity and books such as the Bible, the Qur'an and the Talmud all contain descriptive passages about jewels. In the Book of Revelations the holy city of Jerusalem was described, 'and her light was like unto a stone most precious, even like a jasper stone, clear as a crystal'. Further on in the text it says,

> and the foundations of the wall of the city were garnished with all manner of precious stones. The

first foundation was jasper; the second sapphire;
the third a chalcedony; the fourth an emerald; the
fifth sardonyx; the sixth sardius; the seventh
chrysolite; the eighth beryl; the ninth a topaz; the
tenth a chrysoprase; the eleventh a jacynth; the
twelfth an emethyst.

In the twelfth and thirteenth centuries the church developed
a sanctifying ritual to cleanse gem stones of sins. After
wrapping the stones in linen the priest would say a prayer that
concluded with, 'Bless the effects of virtue thou hast given
them (the stones) each according to their kind, that whosoever
should wear them may feel the presence of their power and be
worthy to receive the protection of their power. Thanks be to
God.'

The engraving of stones probably started in southern
Mesopotamia and became a highly developed art. Carved or
polished stones usually contained a character or word of power.
The Babylonians believed that they could use magical symbols
engraved on bloodstones to foretell the future. The Norsemen
used the Runes as an Oracle. The Runes were stones inscribed
with alphabetic script, each of whose letters possessed a mean-
ingful symbol and sound. They are still used today as a means
of divination.

Mountains were also believed to be sacred as the great monu-
ments of the mineral kingdom. Some cultures considered them
to be the homes of deity. The hills on either side of the Valley
of the Kings in Egypt are alive with gods, the atmosphere
frequently more electrifying than that of the pyramids.
Mountain tops were associated with the gods of sun, rain and
thunder. Sacrifices were sometimes made to appease these
gods.

Mountains were considered to hold in balance spiritual
energies that would be made available to those who climbed
the top. Messages were often received in exactly the same way
as when consulting an oracle. Pilgrimages up a sacred mountain — especially in the Japanese and Chinese, Tibetan and
Indian traditions — meant the renunciation of worldly desires
in order to follow a path of spiritual evolution. These journeys
were often inscribed on jade tablets.

To collect pieces of stone from magical mountains such as the Himalayas was supposed to give one the key to the door of Shambhala, the mystical city of light where eternal youth and perfection hold sway forever. Shambhala was originally a Tibetan tradition but has been adopted by other cultures.

I was told there was a door to Shambhala in Mt Shasta, California, so I decided to climb it. I was also told that if you were not meant to reach the top the mountain itself would prevent you from doing so. I fell over so many times that I felt as if I were being pushed by a giant and invisible hand. It became dark and so I spent the night on the mountain, black and blue, only half way up and without ever finding the door to Shambhala. However I did receive a lot of information that showed me what I needed to do next in my life. A stone came home with me to remind me of the experience.

Much later while climbing in the Sacred Valley of the Rila Mountains, Bulgaria, I was shown another door to Shambhala. The legend there was virtually the same. 'Beings of light', particularly those of the White Brotherhood − an angelic force that guides and teaches humanity − are believed to reside in these mountains. People who believe in the Brotherhood are taught that once a year it is beneficial for their spiritual health to go and stay high up in these mountains for as long as possible.

The Bulgarians told me that the vibration of these light beings is so fine that they are happier in the more rarefied air of the mountains. One discipline of these people is a daily climb to greet the sun − the sun being a manifestation of God. They recognize how important it is to revere water, whether it be a spring, a pool or a pond. White stones and crystals are considered to be sacred. These white stones are used to clear any area containing water to ensure its free flow. I visit the Rila Mountains once a year as part of my own pilgrimage and spiritual renewal. I have seen faces moving among the stones and heard voices and singing that appear to come out of the mountain itself.

I have not found the city of Shambhala but I have experienced great beings of light cavorting about in the mountains as they pour into the earth love, power and light. Time and space, spirit and matter, seemed to meet there and I felt every cell in my body revitalized.

One of the prayers taught to the Bulgarians by their leader and teacher, Peter Deunov, is this one:

> The Disciple must have a heart as pure as a
>   crystal.
> A mind as bright as the sun.
> A soul as vast as the Universe
> And a spirit as powerful as God and one with
>   God.

In common with many other traditions the Bulgarians also have a Sphinx to which they climb and put questions. It is a massive slab of granite rock that sits impassively looking out across a vast valley. Like the Sphinx in Egypt and another Sphinx which I've seen in the Drakensburg Mountains of Natal, South Africa, it appears to contain ageless wisdom. My own questions have always been answered.

The stones of Iona and Findhorn Bay in Scotland have, for me, the same power. The Iona stones seem to pulse with the energy of St Columbus whose presence sanctified the island. Because the island is known for its marble, many of the stones are marble too and reflect beautiful and varied shades.

Stones that I have invited home with me, from any part of the world — and sometimes they refuse the invitation — appear to store a connection with their source area just as one stores in one's mind the telephone number of a friend in another city. One can pick up the stone or dial the number and instantly feel reconnected.

In all nomadic and hunting tribes rocks and stones were the bones of Mother Earth, the mineral kingdom was the point of communion between God and man, and the crystal was the aristocrat of this kingdom. It is still considered the most evolved or developed stone. Crystals reflect the most light and ancient priests believed them to be a God-given force that defies all evil.

For the Aboriginals of Australia crystals are solidified light that contains the Great Spirit. Theirs is a mystical system that uses the landscape and dreaming as its sources of inspiration. To go 'walkabout' is to go on a journey that follows the paths of the ancestors. Crystals are an important tool for the journey, being used to enhance concentration and generate energy.

The Aboriginals still use crystals for medicine, sometimes sewing them just under the skin. They particularly value the Rainbow crystals containing colour prisms because these hold the energy of the Rainbow Serpent — the Rainbow being the bridge between the two worlds — and so have extra power.

An Aboriginal tribal elder once told me that a crystal that comes into one's life by itself, spontaneously and un-expectedly, has grown itself since the beginning of time especially for one. It is therefore like an extension of one's own body and spirit and should be especially valued. This applies to any crystal that appears almost magically — i.e. which you have not bought.

I myself have certainly found that the crystals that have appeared in my life by themselves have had a particular and powerful effect on me. I once picked up a crystal growing out of a stone from the roadside during a visit to Cape Point in South Africa — it was under a rainbow too. Another was a surprise gift from a friend. This crystal had been found in a pyramid in Egypt and when I put it on top of my head I felt as if my body would dissolve into thin air. I discovered that these crystals have a mind of their own and will often disappear for weeks or even months and just when I think they have gone forever they will quietly reappear.

The Cherokee Indians consider the crystal to be the most sacred and precious stone for healing. The Apache medicine men believed that crystals can induce visions and help them find stolen ponies.

In Greece crystals were used to light the sacrificial fire. I have heard that when Marcel Vogel, the great crystallographer, strokes a crystal today sometimes a flame will appear at one end! In Japan, the smaller, opaque rock crystals were believed to be the congealed breath of the White Dragon while the larger and more brilliant were said to be the saliva of the Violet Dragon. The dragon symbolized for the Japanese the highest powers of creation.

The Cherokee Indians still revere their crystal skulls known as singing or speaking skulls because their jaws move. These crystal skulls were also used by the Mayans, Aztecs, Ancient Egyptians and the Tibetans (see page 27). They were also placed on altars and used by priests as oracles. The word of God was believed to come through them. One of these skulls is in the British Museum. Another is in Paris. The Mitchell-Hedges skull was found in an ancient Mayan temple and is supposedly 20,000–50,000 years old. Many psychics consider these skulls to be memory banks recording information in the same way that we store memory in computers today.

An ancient Mexican crystal skull

Because crystals grow with a North and South energy field a researcher developed the theory that early Norsemen used them as an aid in a simple navigational system. When a crystal is rotated so that the poles line up a pulsation occurs. When out of line the pulsation ceases. The same principle is used today in hydrometers — an instrument which determines specific gravities of liquids.

Through the ages almost every 'Wisdom Teaching' or spiritual teaching contains information about the use and abuse of the mineral kingdom. From simple primitive people who loved the colour and shape of stones, to the priests and magicians who practised occult art or the shamans and wise men, who used stones for healing, precious gems have always exercised a magical influence over humanity.

Against this tapestry of history what exactly are stones, gems and crystals?

The secret life of crystals

I used to wish I was a rock or a stone, a part of the mineral kingdom. I thought, 'It needs no other form of life to support it. It is not dependent on anyone else's approval or disapproval, it simply is.' I did not realize then how true this was. I now know that minerals, as well as having wonderful shapes and colours, were formed by natural processes which have not been directly affected by living things.

Minerals are part of the crust, and probably the interior, of the earth on which we live. They are formed out of molten rock called magma which rises from deep within the earth and solidifies on the surface. Gradually, affected by the weather, and ecological disturbance shifting the earth's crust, this rock is pushed back into the depths and melts once more. Gemstones are products of this vast geological recycling process.

Crystals and gems grow in cavities, usually far below the earth's surface, when tongues of molten rock (magma) intrude into solid rock, such as granite, and cool slowly. Magma that erupts, such as in a volcano, cools too fast to allow crystallization to occur.

The cavities in which crystals grow are called pegmatites. Pegmatites provide the space and chemical isolation necessary for the formation of pure mineral crystals. Geologists and miners sometimes call them nature's jewel boxes.

Different types of magma create different stones. The chemicals, the texture and density all play a part in the gemstone that will form. A magma rich in iron and magnesium — gabbro — will create peridot, zircons and sapphires. Diamonds are found in magma known as kimberlite and are really crystallized forms of pure carbon. Quartz crystals are formed from magma rich in silicon and oxygen, and start their growth from silicate-type seeds attached to the cavity in which they grow.

In contrast, Herkimer diamonds (from Herkimer County, New York State) — which are natural gemstones that appear to have been faceted and polished — grow floating in a liquid. Amethysts are often found in cavities left by gas bubbles in solidified lava.

No one seems to know exactly how long it takes crystals to grow. Because of the ancient belief that crystals were solidified light, or holy water poured down by the gods, many people now believe they are millions of years old and part of the foundation of the world. A friend of mine asked an old man sitting outside a crystal mine in Arkansas, 'How long does it take a crystal to grow?' He said 'I've been sitting here for sixty years and that's certainly not long enough to find out.' I myself asked a geologist the same question. He said 'Nobody knows.' However in a laboratory a crystal can be grown in a matter of weeks.

Because crystals are part of the great life-cycle of the Earth — the molten rock rising and sinking, forming and re-forming, we probably do have available to us crystals and gems that could be a million years old and others that are relatively young. For me it is unimportant. My relationship with a crystal is based upon how I feel about it personally. The effect it has on me has nothing to do with age.

Crystals come in many sizes. The ways they are mined are equally varied. Crystals and gems growing in veins or pegmatites are sometimes mined by native miners using simple hand-tools. More complex mining methods using drills, explosives and bulldozers are employed in huge crystal caves.

In Brazil it is often necessary to follow a vein for several feet before it opens into a cavity where clusters of crystals have formed. In Arkansas there are sandstone caves which, when dug out, reveal vast sheets of crystals. Each sheet contains thousands of crystals individually more than a foot long.

Many precious stones that have been washed into river beds — for example in Sri Lanka — are picked out by washing the gravel through sieves. Twenty years ago in South West Africa it was possible to pick up chunks of quartz, amethyst, citrine, and many other fabulous stones, as one walked around. Although this can still be done today the quality of stone is not the same. I have picked up clusters of quartz in the mountains of Bulgaria. They were fairly opaque but still extremely beautiful.

I am often asked if the vibration of a crystal is affected by the area in which it grows or the way in which it has been mined. For example for me a crystal from Brazil has a totally different 'feel' to a crystal from Arkansas. I have an affinity with Arkansas crystals — I get a buzz from them, whereas the stones I have collected from Brazil are beautiful but distant. It is as if I am with a very beautiful woman who has no warmth or personality.

This does not mean Brazilian crystals are in some way negative. Many of my friends find them warm and endearing! Therefore, yes, the area in which a crystal grows is bound to affect its vibration, just as tomatoes grown in Australia have a different flavour to tomatoes from Guernsey. To find out which tomato you prefer you have to taste it.

Every country has a different 'feel', a different energy, as well as the difference in the chemical make-up of the soil. We are usually more comfortable in one area rather than another. People have a different feel too, a different 'vibe'. We even talk about being on the same wavelength. It is exactly the same with gems and stones. You must find the ones that vibrate to your own rhythm or for your own needs.

Mining also affects the crystal. When sheets of crystals are scooped away from cavern walls by bulldozers they tumble down and are sometimes damaged. Clusters may split apart and cause individual crystals to bear a fracture or mark. When crystals are prized apart to separate them for sale, they may also become chipped or cracked.

If one considers what a stone goes through during its journey from deep within the earth to your bedside table, it is amazing how many crystals remain unflawed. However a crack in a crystal can be a source of delight as it will usually create wonderful patterns that reflect colour. As with a person it is often the flaw that creates individuality and beauty.

The sharp angular point of natural quartz crystal, called a

termination, has not been shaped by man. Crystals come out of the ground naturally faceted. The faces or facets, and most crystals have six, are formed by the density of the atoms within. In other words the shape of a crystal is determined by its internal molecular structure.

Crystals consist of millions of individual structural units of atoms called unit cells. The unit cell is a square with an atom at each corner. Crystals are atoms composed of silicon and oxygen linked together in a network or pattern called a lattice. The effect of heat and pressure on these atoms rearranges them and this is known as crystallization.

Quartz crystals, when removed from the earth, react when squeezed or pressed. They produce a current of electricity in a process called piezoelectricity. This piezoelectricity or oscillation is why crystals are used in radio, television, computers and in any electronic device that demands high precision.

The application of heat to natural quartz changes the negative and positive charges at either end of the crystal. The temperature change disturbs the stability of the atomic structure. Attempts to regain stability produce opposite charges at either pole. This is called pyroelectricity.

If pressure and change of temperature affect a crystal then bashing or prizing it out of the earth through mining must do the same. However, it is the nature of a crystal to balance and harmonize. It does this by re-arranging the atoms of its internal structure. By doing this it can, when the external application of energy or force stops, return to a point of equilibrium.

In spite of these internal mechanics I sometimes sense shock in a crystal and I use the power of love and thought to heal it. Love is a measurable force. So is thought. Energy follows thought exactly like a current of electricity. The crystal responds to these currents in the same way as it does to other energies.

Some crystals are more naturally receptive than others. They may record the memory of the attitude of the people who have handled them as well as the memory of being yanked out of the earth. If your crystal appears to be traumatized you can heal it by loving it. Hold it, stroke it, talk to it — just as you might behave with a shocked child.

Healing damaged crystals

Two or three years ago I acquired a crystal that was so damaged it seemed to be one massive web of fractures and broken edges. During a crystal workshop I passed it around the room, to approximately 80 people, who kept their eyes closed. I asked them to write down what they had experienced as they held it. Each person used words such as 'fear, horror, broken-hearted, shattered, smashed, destroyed,' to describe their feelings. I then passed the crystal around again and asked the group to heal it.

A week later, when this same crystal was passed around among a different group of people the words used to describe the crystal were like joy — 'pleasure, purring, delight, freedom, happy'. The crystal had been healed.

I finally gave this crystal to a woman who was emotionally very unbalanced. I asked her to complete the healing. As she did this the crystal not only healed her in return but grew a little rainbow as if to confirm, 'I'm better.'

I also met a woman who had been given a beautiful piece of semi-precious jewellery by her husband for Christmas. Each time she wore it she felt uncomfortable and became ill. She asked me if I believed a gem could be affected by the thoughts of the jeweller who cut it. I told her that we project, consciously and unconsciously joy, sadness, anger or boredom into everything we do, whether it is shaping gems or preparing food. We cleansed her necklace of any negativity and she now wears it happily. We later discovered that the jeweller's wife had run away with his business partner. His despair and anger had gone into the jewellery.

We can cleanse, heal and transform stones and jewels by projecting our breath, empowered with love, into them. I will describe this in Chapter 5 as it is also an important technique in cleansing crystals.

Famous and favourite gems and stones

Jewels as objects of beauty and rarity have always been highly prized and valued. For thousands of years they have been used as amulets and talismans as well as decoration for almost every part of the human body.

Stories associated with their transfer from owner to owner through the ages cover an extraordinary panorama of characters. Kings, queens, soldiers, politicians, princes, paupers, smugglers, swindlers have all been seduced by the lure of gems. Truth and fantasy have become entwined in legends that span some of the most dramatic periods in history.

Writing about diamonds in the 1890s Gordon Williams, the manager of De Beers, said

> To win them temples have been profaned, palaces
> looted, thrones torn to fragments, princes
> tortured, women strangled, guests poisoned by
> their hosts and slaves disembowelled. Some have
> fallen on battlefields, to be picked up by ignorant
> freebooters, and sold for a few silver coins. Others
> have been cast into ditches by thieves or
> swallowed by guards, or sunk in shipwrecks, or
> broken to powder in moments of frenzy.

This wonderful statement applies equally well to all jewels. Murder, intrigue, romance and robbery all play their part in stories more adventurous than any tales from the Arabian Nights. If jewels could speak, what fascinating truths we would learn about their history!

Many writers have been inspired by gems and precious stones. Walter de la Mare wrote:

> Ruby, amethyst, emerald, diamond,
> Sapphire, sardonyx, fiery-eyed carbuncle,
> Jacynth, jasper, crystal-a-sheen;
> Topaz, turquoise, tourmaline, opal,
> Beryl, onyx and aquamarine;
> Marvel, O mortal! — their hue, lustre, loveliness,
> Pure as a flower when its petals unfurl —
> Peach red cornelian, apple green chrysoprase,
> Amber and coral and orient pearl!

De la Mare was not the only writer to use the visual power of gems in his creations. Shelley, Shakespeare, Wilde, Maupassant and many other writers have employed jewels as a basis for plays, poems and stories to excite and entertain. One of the most important works in Middle English is the long poem *Pearl* which likens a father's loss of his daughter with the loss of a pearl and uses the jewel to symbolize the human soul.

In the first century AD the Roman author Pliny the Elder, writing about diamonds, said that they were so hard that when hit even the anvil broke. He added however 'this invincible element can be broken by ram's blood. The blood must be fresh and warm and even so many blows are needed.'

The names of precious and semi-precious stones are frequently used as adjectives to describe physical or spiritual beauty. For example, pearly teeth — ruby lips, sapphire seas, diamond eyes, pearly skin, clear as a crystal. Shelley wrote about 'the crystalline sea', 'an emerald sky' and 'the chrysolite of sunrise', and Shakespeare wrote 'those are pearls that were his eyes'.

During much of its history jewellery has been worn as a sign of social rank as well as a talisman to avert evil and bring good luck. Generally any stone if perfect was considered a source of blessing while a misshapen stone, or one that lacked lustre, was thought to bring misfortune.

Gradually popular beliefs began to associate particular qualities with certain stones and colours. Red was considered revitalizing and cleansing for the blood, yellow for problems with the liver, green for soothing the eyes, blue to clear the mind and lift the spirits.

Later this developed into engraving the stone which was then considered to be an even more powerful source of protection. As astrology became a sophisticated science it was believed that this protective power could be increased if the engraving was done at certain phases of the sun and moon. Sometimes this was aligned to the hour and date of birth of the wearer which increased the amulet's effectiveness.

There was a difference in the talismanic wearing of stones to prevent disease or misfortune and the use of them as a medicine. The diamond named from the word 'Adamas' meaning unconquerable when worn or held next to the body brought strength and invincibility. If taken internally it was considered to be a dangerous poison. The Hindus thought that the powder of a flawless diamond would bring long life and happiness while that of a flawed diamond would bring death by poison.

The death of Emperor Frederick II of Germany in 1250 was blamed on powdered diamond as was that of the Turkish Sultan Bejazet in 1512. Catherine de Medici was supposed to have used powdered diamonds in conjunction with the poisons she administered to her victims in the sixteenth century.

The discovery of diamonds is attributed to Alexander the Great around 350 BC. There is a legend about the 'Valley of the Diamonds' being guarded by snakes whose mere gaze could kill a man. Alexander instructed his soldiers to polish their shields to a mirror-like finish before advancing on the snakes. With their image turned on themselves it was the snakes and not the soldiers who died. Alexander's men then threw greasy carcasses of sheep into the valley. The diamonds stuck to the carcasses but then eagles, attracted by the meat, swooped down and carried them off. The story says that Alexander's bowmen were able to shoot the birds down and recover the diamonds.

An Indian legend says that diamonds were cursed by a goddess from whose statue the diamond eyes were stolen. As a punishment she decreed that diamonds would bring misfortune to anyone who owned them.

Stories associated with famous gems

The story of the Hope diamond would certainly seem to support this Indian legend. Brought from India for Louis XIV in 1668, the King had it cut for his mistress Madame de Montespan and it became known as the French Blue. It was later worn by Marie Antoinette. After the Revolution it was stolen and disappeared until 1830 when it was bought by Henry Phillip Hope. In 1906 it passed from the Hope family into the hands of a Persian gem dealer, Jacques Celot, who committed suicide. It was then bought by a Russian, Prince Kanitovski, who gave the diamond to his French mistress to wear for her performance at the Folies-Bergère. He shot her dead in the middle of her act. Two days later Russian revolutionaries stabbed the Prince to death.

The next owner, an Egyptian, drowned with his whole family in a steamer collision off Singapore. The broker who handled its next sale, to the Sultan of Turkey, was killed, with his wife and child, when he drove his car off a precipice. The Sultan gave the jewel to his mistress but during a revolt by his army shot her dead. He was exiled. The eunuch in charge of the jewels was hanged.

Cartiers then bought the Hope diamond and sold it to the Maclean family. The Maclean's eight-year-old son was run over and killed by a car. One of their daughters and a grand-daughter both died of an overdose of barbiturates. Edward Maclean had a nervous breakdown and eventually died in a mental hospital.

Eventually Harry Winston, a New York gem dealer, bought the Hope diamond and presented it to the Smithsonian Institute in Washington. The curse is now supposed to have stopped. However, among the thousands of letters thanking him for his donation, one begged Harry Winston to take the diamond back. 'The country has gone to pieces since its arrival at the Smithsonian', the letter said.

An African king, King Lobengula, also had his life ruined, after acquiring biscuit tins filled with smuggled diamonds from the Kimberley mines. Lobengula was a huge man known as 'Drinker of Blood, Calf of the Black Cow, Man-eater and Lion'. Betrayed by the mining company and by Rhodes, who used the disturbance as a pretext for war, Lobengula fled. He and his three sons died of smallpox soon afterwards. His biscuit tins of diamonds, believed to be worth five million pounds, vanished and have been sought by fortune hunters for decades.

One of the most brazen confidence tricks in history which involved diamonds and contributed to the downfall of Marie Antoinette during the French Revolution began in 1772. Madame du Barry, mistress of the ageing Louis XV, demanded that he should buy her the most expensive diamond necklace in the world.

Louis commissioned the court jeweller to search Europe for the finest diamonds he could find. Bohmer, the jeweller, bought approximately 600 which he strung together in a magnificent necklace. Before he could deliver this necklace the king died of smallpox. Neither the new king, Louis XVI, nor Marie Antoinette, his twenty-year-old wife, had any use for a necklace, 'vulgar like a scarf'.

Marie Antoinette's distaste for the necklace was exceeded only by her loathing for Cardinal de Rohan who had been French Ambassador to her mother, the Empress of Austria. The Cardinal was well aware of the Queen's dislike and wanted, more than anything else, to ingratiate himself with her.

At this point a self-styled countess, Jeanne de la Motte, entered the story. She convinced Cardinal Rohan that she had great influence with the Queen and could restore him to royal favour. With the aid of a forger she produced a number of letters, apparently in Marie Antoinette's handwriting, suggesting that the Queen had softened towards the Cardinal.

She then arranged a meeting, in the dark, in the gardens of Versailles between the Cardinal and a girl who somewhat resembled the Queen. The Cardinal was overjoyed. Not only had the Queen forgiven him but she was apparently in love with him too. When he received a note asking him to buy for her the previously scorned diamond necklace he was delighted to comply. The transaction would be secret and the go-between would be Countess de la Motte. The Cardinal bought the necklace, passed it to Jeanne de la Motte who gave it to her husband to bring to England, where it was immediately broken up and sold.

It was about six months before the Cardinal plucked up the courage to ask the Queen why she did not wear the necklace. Marie Antoinette was not only outraged but brought the matter before parliament. Jeanne de la Motte was arrested, found guilty and thrown into prison from which she later escaped. Marie Antoinette had never been popular and so the mob was delighted to believe that the Cardinal had been her lover and bought her rich jewels while they starved. This episode is considered to have been one of the sparks that set alight the French Revolution. Marie Antoinette was guillotined in 1793 and Jeanne de la Motte died while jumping out of a window in an attempt to avoid her many creditors.

The most recent diamond to cause international dispute is the oldest and most famous gem in the British Crown jewels – the Koh-i-Noor diamond. In 1976 the Pakistan government demanded its return saying that it had been illegally removed. The Koh-i-Noor has been fought over in many bloody battles since 1304 when it belonged to the moguls of Malwa. There is a legend saying that he who owns the Koh-i-Noor rules the world. In 1833 the Afghan prince, Shah Shuja, was blinded and tortured for days before he would give up the Koh-i-Noor. When asked to explain his resistance he said, 'It brings good luck and has ever been the bosom companion of him who triumphs over his enemies.'

The Koh-i-Noor was named by Nadir Shah, leader of the Persian army who invaded India in 1739. The defeated mogul emperor, Mohammed Shah, had hidden the stone in his turban. Told of this, Nadir Shah invited him to a feast and, observing the ancient custom, proposed an exchange of turbans. Mohammed was in no position to refuse. Once Nadir had Mohammed's turban he unrolled it and on seeing the flashing brilliance of the diamond exclaimed, 'Koh-i-Noor' – meaning Mountain of Light. It was brought to London in 1850 and presented to Queen Victoria, who had it recut. It became the centrepiece of a new crown for the 1937 coronation of Queen Elizabeth and George VI.

The biggest diamond ever discovered – the Cullinan diamond – is also a part of the Crown Jewels. It was presented to King Edward VII in 1907 and cut into nine major gems. It was said that when the stone was cut a doctor and a nurse were in attendance (because it was such a stressful business!) and that Joseph Asscher, the cutter, fainted as soon as the work was done. The stones not used in the Crown Jewels are in the personal possession of the Royal Family.

The Prince's Ruby, now the centrepiece of the Crown of State, was brought to England by Edward the Black Prince. Edward agreed to help a fourteenth-century Spanish king known as Pedro the Cruel, who had been banished from his own country, to regain his throne. Pedro promised payment in money and lands. After Edward had duly restored Pedro to his throne he was given in payment the red ruby. Only many centuries later was it discovered that it was not a ruby at all but a chunk of spinel. Spinel is similar in colour to a ruby but worth only a fraction in value.

Another stone with a background of violence is the Regent diamond now in the Louvre, in Paris. In 1722, it was regarded as the most important jewel in the French treasury. It was carried by Napoleon in the hilt of his sword when he was crowned Emperor.

Napoleon's carnelian seal

Napoleon was also superstitious and carried about with him an octagonal-shaped carnelian which he had found in Egypt. He left an injunction for his son that said, 'I desire that he will keep as a talisman the seal which I used to wear attached to my watch.' It was engraved with the words, 'The slave Abraham relying upon the Merciful (God)'. The Regent diamond had been found in India by a slave who smuggled it out of the mine concealed in a self-inflicted wound on his leg. He then made for the coast and offered half the value of the stone to a British sea-captain for safe passage out of India. The captain agreed but at sea he killed the slave, took the diamond and flung the body overboard. He sold the stone but is supposed to have squandered the money in bars and brothels and eventually hanged himself.

In the past the greed that has led a man to risk everything for the sake of a jewel has usually caused his death, disgrace or punishment. A strange exception is the case of Captain Blood who, in 1671, dressed up as a clergyman and stole the Crown Jewels from the Tower of London. Despite being caught he was not only pardoned but given an estate in Ireland and a pension of 500 pounds by Charles II.

As the great majority of famous gems have brought misfortune rather than happiness perhaps the ancient Persians were right. They considered precious stones to be the source of much sin and sorrow and believed they were created by the devil to exploit the covetous side of human nature. However that did not prevent their lavish use of rubies and sapphires for decoration.

Amulets

With the linking of certain energies and qualities to particular precious and semi-precious stones, various forms of jewellery have been used throughout history as much for protection as decoration. In fact it is likely that man thought of adorning his body with shells, bones, teeth and pebbles long before he thought of wearing clothes. Prehistoric burial grounds have revealed necklaces of lion and tiger teeth, presumably worn as much for the power of the animal as for decoration. Pictures of prehistoric African women found painted on the walls of caves have shown surprisingly ornate jewellery.

In India, Sri Lanka, Thailand and Burma the wall paintings show lavish use of gems. The clothing of these women seems to have consisted entirely of such jewellery as necklaces, earrings, amulets, bracelets, toe-rings, ornate belts and 'cache-sex' — serving the same function as a loin cloth.

The most ancient examples of jewellery were probably found in Queen Pu-Abis's tomb at Ur in Sumeria, dating from the third millennium BC. The Queen's body was covered with a robe made of gold studded with lapis lazuli, carnelian, agate and chalcedony beads. The practice of burying the dead with precious jewels to aid their passage into the next world also made use of talismans to protect them from evil spirits. The Egyptians considered the shape of an eye made from lapis lazuli — lapis signifying truth and mental clarity — to be an amulet of great power. A winged scarab was also frequently placed in a tomb. The scarab — a carved dung beetle — was the symbol of the sun-god: the dung beetle lived underground and took flight at noon, thus becoming a symbol of reincarnation.

Scarabs were used as seals and worn as talismans. They symbolized immortality, rebirth, regeneration, the blend of left and right brain and the initiation to a higher level of consciousness. The lay-out of many of the temples followed the design of the scarab and like the Mandalas of India and Tibet illustrated a way whereby man could reach enlightenment. Scarab amulets were often made from amethysts. The green emerald, a precious stone for the Egyptians, was supposed to bring fame and fortune and, if sucked, to give the gift of prophecy. In the case of marital infidelity it broke, however.

In Greece engraved talismans covered a variety of subjects. In the fourth and fifth centuries designs of women bathing and the body beautiful were greatly admired. In the mid-fifth century the scarab, with the same symbolic significance as it had for the Egyptians, became popular. By the sixth or seventh centuries jewellery usually showed deities and mythological figures. The Hercules knot was considered a magic knot and in jewels it took on the significance of an amulet. It was thought one could draw the power of the gods around one by wearing their image.

In Rome Etruscan scarabs were popular until the third century. Later gem collecting became a passionate pursuit. Julius Caesar was an avid collector and is said to have given six separate collections to the Temple of Venus Genetrix. The most common stones collected were garnets, carnelians, amethysts, topaz and peridot. Red jasper and carnelian were believed to have magical qualities. To drink out of an amethyst goblet was believed to be a protection against drunkenness.

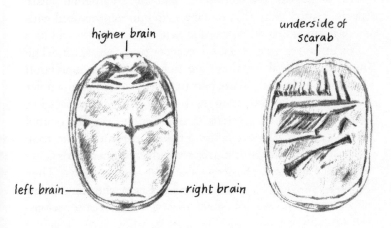

higher brain

underside of scarab

left brain — right brain

A *scarab*

This belief came from a legend that Bacchus in a drunken rage had commanded tigers to eat the first human they came across. This happened to be a a young girl, Amethyst, who was worhipping at the shrine of Diana. In terror she begged Diana for help and was turned into a statue of white crystal. Bacchus was ashamed, calmed down and poured grape juice over the statue, turning it into the colour now called amethyst.

Brown coral was supposed to delight evil spirits and oriental mystics also warned against the wearing of discoloured coral. Plato suggested that coral was good for children both to wear and rub on their gums. Presumably he meant the correct colour.

Jewellery surviving from the Ming dynasty again shows the use of more semi-precious stones than actual jewels — amethyst, agate, chalcedony, jade and many varieties of quartz. Jade, especially white jade, was the most admired stone and symbolized the stone of heaven — the stone of divinity.

The Chinese mandarins wore different stones to denote different ranks. The first rank wore red, a ruby or red tourmaline, the second wore coral or garnet, the third blue, beryl or lapis, the fourth rock crystal and the fifth, 'any white stone'. The garnet has always been considered a lucky and revitalizing stone. Not only was it used in amulets but the Hunzas in Kashmir used garnets as bullets believing they would cause more harm. Their enemies were delighted!

Not only are jewels themselves a continual source of delight but the stories about them are rich, fascinating and neverending. From the Indians burning gold and emeralds before images of the sun and moon, to the gloves of St Martial, which were studded with precious stones that sprang out of their settings if a sacrilegious act had been committed, tales about gems are full of magic and mystery and cover the whole gamut of human fantasy and experience.

Pliny said, 'Some gems are considered beyond price and even beyond human estimation so that to many men one gem suffices for the contemplation of all nature.'

In the middle ages a ruby ring was thought to protect its owner against seduction while drawing into his life wealth, lands and titles — but only if worn on the left hand.

The Maoris regard the green jade tiki as being a lucky stone. The tiki was originally called the hei-tiki, meaning a carved decoration for the neck. It was passed from one owner to another on death and was believed to contain and impart the wisdom of the ancestors. The tiki has been a lucky stone for me too. Whenever I've worn it, especially in New Zealand, I have had happy adventures — such as bumping into people I haven't seen for twelve years or being unexpectedly invited to participate in a Morae — or meeting — with the Maori people. It has for me been a warm 'people stone'. Another time I was wearing it on a flight to New Zealand and I suddenly found myself being ushered into a first-class seat thickly lined with sheepskin. It was a fantastic and luxurious way to fly and completely unexpected.

Some of the good or bad luck seems to be associated with the brilliance of the gem. When the lustre dims in any stone it is generally believed to be a bad omen foretelling sickness or death. Bishop Constantia wrote about the stones on the vestment of the high priest assuming 'a dusky hue or becoming the colour of blood' if the people had sinned. If the stones shone 'like the driven snow' it was cause for celebration. Another writer speaks of the breastplate of the high priest being used to assess guilt or innocence in cases of theft. If the person under suspicion was innocent the stones glowed, if guilty they became dim.

The belief in particular qualities for different stones and colours seems to coincide in most cultures, although occasionally there is a total contradiction. For example the onyx, for some, is a stone to balance, or 'earth and ground' (anchor) one. However, the Arabic name for onyx — el Jaza — means sadness. In Arab countries it was thought that the contrasting black and white bands signified separation and disagreement — although beneficial to ladies about to give birth!

The history of the owner of a particular piece of jewellery will also impart beliefs in the fortune or misfortune of a stone. However, all stones carry a vibration — more than that, a stone is a manifestation of a certain vibratory level of energy. Some stones obviously resonate with a corresponding vibratory note within us. This will vary from person to person in the same way as a fruit can be 'good' or 'bad' for us, depending on the acidic balance of the body. Therefore wearing jewellery with the correct vibratory note will stimulate balance.

If this note does not correspond or resonate with one's own, it can create imbalance and a sense of being out of step. This in turn will be more likely to attract misfortune or bad luck.

Paramahansa Yogananda, the Indian mystic and yogi, was told by his master, Sri Yukteswar, to wear an amulet of silver and lead to counteract impending liver trouble. Sri Yukteswar had enormous understanding of universal forces. He also said 'Electrical and magnetic radiations are ceaselessly circulating in the universe. They affect man's body for good and ill — pure metals counteract the negative pulls of the planet. *Most effective of all are faultless jewels.*' He meant that precious jewels and pure metals have a protective effect for the wearer.

A friend gave me a silver coin which had been picked up in a field in Belgium by her father over fifty years ago. It had such a strong vibration I could feel it emanate from within her hand before I myself had even touched it. When I put it around my neck the cells and bones of my body moved into place as if a chiropractor had given me a spinal adjustment. However when she wore it it would not stay on her neck. Either the coin fell off or the chain broke — this happened a number of times — or the coin disappeared for days as if sulking. She finally realized that the coin was not for her. For me it still imparts strength and balance and acts as an amulet of power. Incidentally, aside from the natural vibration of metal or stone we automatically empower the things we value. We impregnate what we wear with the power of our thoughts. Energy, we know, follows thought and energy affects all matter.

I had an uncomfortable experience of this with two different gems. One was in a ring. The previous owner had impregnated such a heavy, depressed vibe into this ring that everyone who held it was affected. As I mentally cleaned it, and I needed the help of two other people to do this, it became lighter and literally weighed less. We measured it before and after on scales. The heavy negativity had made the ring feel heavy.

The other gem was an aquamarine I acquired from a man who looked very much like a magician. The stone behaved as if it had been programmed to bleep out, at regular intervals, a message for me to call this man. It was like a psychic telephone and brought an extremely manipulative power into the room when it 'rang'. As soon as I cleaned it the aquamarine became light and glowing and grew a tiny rainbow. I cleaned it with water, breath, visualization and love.

Another favourite possession of mine is a small green marble heart. Originally it had inside it a perfectly formed white heart shaped by the patterning of the marble. During a visit to Sri Lanka I experienced a form of psychic attack and the white heart within the green marble heart vanished. The marble absorbed the negativity into itself to the point that its own heart dissolved. It was extremely distressing and despite my giving it love the white heart has never returned.

I have seen a similar incident occur with a stone that absorbed a friend's illness. Again it not only changed colour but became heavier. A French woman writer said of her own jewels that they became pale if she neglected them. She used to wear all her rings at the same time because she said they became jealous of one another if one gem was left behind.

The shapes and symbols of jewellery can also affect us. We are all bombarded by the symbols around us. Most of us are totally unaware of their effect on us.

A friend of mine in Australia was bemoaning the fact that there was no man in her life. I pointed out that every day she wore around her neck a locked locket in the shape of a heart. I suggested that she take it off and unlock it and put it on her bedside table where she could look at it before going to sleep and impress upon herself that her heart was now open. Within about ten days she met a man and began a relationship which flourished.

Look at the symbols you wear and surround yourself with. What do they say? If they are helpful, talismanic, keep them. If not, change them or give them away.

How to choose a precious or semi-precious stone

Cleopatra, one of history's most exquisite beauties, took the greatest care in selecting the wide variety of precious gems used to enhance her loveliness. Her bejewelled headdress was created from gems believed to engender the qualities of magnetism, youth and vitality. Her jewellery was chosen to ensure inner, as well as outer health, and perfection.

The Ancient Egyptians understood the vibratory forces of certain stones and their correlation to similar forces within the body so there was a clear purpose behind their choice of gems. The Egyptian sapphire, which we know as lapis lazuli, was ground up to make a brilliant eye shadow and, when soaked in water, was used as an eyewash to brighten the eyes, just as we use eyedrops today. The chemical composition of lapis lazuli includes copper which is an astringent. Because of this it was probably an effective eye treatment. Blue stones were associated with the sky, heaven and Ra, the Sun god. They were thought to aid chastity by cooling the fevers of sexual passion.

Malachite was another stone much favoured for eyeshadow. Its brilliant green colour was believed to aid good eyesight and stimulate visionary powers. If a piece of malachite broke in two it signified danger. Only recently a friend of mine was holding a malachite egg when it broke. She went home to discover that her supposedly reformed, alcoholic husband had gone on a binge and that her son and daughter-in-law, parents of her favourite grandchild, had decided to separate.

In Ancient Egypt, mother of pearl was rubbed on the skin to make it sparkle, and rubies and garnets were used to stimulate the cells while coral improved the circulation. Emeralds were believed to sharpen the mind and quicken the wits if held under the tongue, and amethysts were used to reduce intoxication and inflammation, while aquamarine was also thought to be good for reducing fluid in the body and topaz was believed to help the kidneys and bladder. For the Egyptians the topaz represented the sun and was used as a symbol of life and fertility. It was meant to banish the terrors of the night, including the fear of death.

Recent excavations in Egypt show that diamonds were used to cut stone as early as 4000 BC — although it is thought that what passed for diamonds worn on the body were in fact quartz crystal. Crystals were believed to be the connecting link between heaven and earth and were often used in statues, to symbolize the all-knowing, all-seeing eye of God. Crystal balls were used for 'scrying', or crystal ball gazing, in order to foretell the future. Much of this quartz appears to have come from crystal mines near Abu Simbel. Occultists believe that the Egyptians built the pyramids by applying the geometric principles governing the internal atomic structure of the crystal.

The use of gems, stones and precious metals for treating the body was well understood by the Egyptians and they refined this into a highly developed science. Colour also played a vital role both in healing the body and as an aid to developing certain powers. One method was to place food in jewel-encrusted bowls of a particular colour in order that the food would absorb the vibration of the colour before being eaten.

I do this myself today. When I sense that my body needs a certain vibration I wear the clothes and eat the food that can provide it. Instead of using jewel-encrusted receptacles, I use plain colour!

Perhaps Cleopatra's day began with a long soak in a pool fashioned from quartz and filled with asses' milk. To keep her body trim and firm, her skin glowing, hand-maidens would have massaged her with rubies, garnets, coral and mother of pearl while rose quartz soothed away wrinkles and softened her complexion. To sharpen her wits for Mark Antony, she would perhaps have sucked an emerald while contemplating the most appropriate colours to surround herself with, depending upon the astrological influences of the day. Enrobed in a gown lavishly decorated with topaz, peridots, crystals and cornelians, her waist encircled with a belt of hematite, her eyes brilliant with malachite and sapphire, she no doubt sipped from an aquamarine studded goblet while listening to the fortunes of the day being pronounced by the crystal scryers.

Another woman who understood and used the magical power of gems was the fabled Queen of Sheba. Black and beautiful, she was the Queen of Africa, and was adored by King Solomon — who sent emissaries to distant lands in search of jewels with which to please her — she chose peridot, onyx and cornelian as her favourite stones.

In much the same way the present-day Cleopatras — Elizabeth Taylor, Shirley MacLaine, Goldie Hawn, Jill Ireland and Ali McGraw — all use gem and crystal power to stimulate their qualities of charm, clarity and emotional stability. Men too, such as Vincent Price, Simon le Bon, Johnny Carson and Richard Gere have also become crystal fans.

Stones of the Calendar

The belief that a particular stone was dedicated to each month of the year seems very ancient and was possibly related to the twelve stones on the breastplates of Jewish high priests. The writings of St Jerome and Josephus comment on a connection between the twelve stones and the twelve months of the year as well as the twelve zodiacal signs.

The Chaldeans* who lived in Mesopotamia around 4000 BC discovered a relationship between the planets and specific gems.

> Saturn was associated with Sapphire
> Jupiter with Jacynth
> Sun with Diamond
> Mars with Ruby
> Venus with Emerald
> Mercury with Agate
> Moon with Selearite

The general wearing of a stone associated with the birthdate

* The Chaldeans were famous for studying the stars to foretell the future.

did not become widely fashionable until the eighteenth century. Before this time stones were used more for therapeutic reasons.

Originally it was thought better to have a stone for each month of the year and wear them in turn. The stone of the month was believed to generate the most healing and protective power at that time. To wear the stone of the sign under which one was born was also meant to draw down and magnify the vibratory influence of that particular planet.

In the sixteenth century it was fashionable to wear an emerald in the spring, a ruby for summer, a sapphire in the autumn and a diamond in the winter.

The emerald is associated with Venus, the goddess of love. Maybe it is an appropriate stone for spring when 'a young man's fancy turns to love'. Its green colour symbolizes growth and sensitivity — new birth. The ruby is known as the king of the jewels stimulating expression and creativity. An ancient Burmese legend tells of a giant eagle soaring over a valley searching for food. Far below he saw what appeared to be red meat so he swooped to earth and tried to seize it. It resisted all his efforts until he approached it with reverence, recognizing that it was a sacred stone, created from the fire and blood of the earth.

Autumn is a sad time for many people. Summer has gone, the leaves are falling, life appears to be winding down. The sapphire is a gem that brings clarity and insight. It protects against depression and helps us to move on instead of looking back.

The diamond too is an appropriate stone for winter. It is cold, hard, brilliant, like the ice and frost on a winter's day in a cold country. It is a hard stone both literally and emotionally. However, it can sharpen the intellect while one awaits the softness of spring.

As the wearing of precious and semi-precious stones became increasingly popular the custom of wearing jewels for different times of the day developed. For example jewels recommended for the night were turquoise, jade and jasper, tourmaline, malachite and lapis lazuli. Jewels for the day were garnet, emerald, diamond, topaz, amethyst, ruby or kunzite. Most of the gems designated for the night are more opaque, as they would help anchor the body, slowing it down while sleeping, whereas those of the day sparkle with light to inspire and stimulate the intellect.

The Industrial Revolution swept away much of the tradition of jewellery being worn as a sign of wealth and status. Gradually more people were able to afford to buy gems and the subsequent popularization of gem collecting led to the custom of associating specific stones with particular wedding anniversaries and the common use of birthstone charts as we know them today.

The first years of marriage seem to be celebrated with the most basic materials, such as paper and cotton. In the following chart twelve years must elapse before a gem enters the picture!

Wedding anniversaries

1 Paper	14 Moss Agate	35 Coral
2 Calico	15 Rock Crystal	39 Tiger's Eye
3 Linen	16 Topaz	40 Ruby
4 Silk	17 Amethyst	45 Alexandrite
5 Wood	18 Garnet	50 Gold
6 Candy	19 Hyacinth	52 Star Ruby
7 Floral	20 China	55 Emerald
8 Leather	23 Sapphire	60 Yellow Diamond
9 Straw	25 Silver	
10 Tin	26 Star Sapphire	65 Sapphire
12 Agate	30 Pearl	67 Sapphire
13 Moonstone		75 White Diamond

An ancient Hindu tradition suggests the following combination of stones to be used as a protective talismanic influence for life — presumably married or single! The ruby, diamond, pearl, coral, jacynth, sapphire, topaz, cat's eye and emerald were associated with particular planets. For example, the topaz with Jupiter, coral with Mars, emerald with Mercury, sapphire with Saturn, diamond with Venus and ruby with the Sun. This does not correlate with some of the modern charts. However it was believed to be an extremely powerful amulet.

Birthstones

After studying many birthstone charts I have compiled a chart which synthesises the essence of such charts all over the world today (but do please note that this does not fully correlate with the ancient Chaldean chart relating planets and gems (see p.50) as this is a modern synthesis).

Capricorn (23 Dec.–22 Jan.)	Ruby, Garnet
Aquarius (23 Jan.–22 Feb.)	Blue Sapphire, Amethyst
Pisces (23 Feb.–22 Mar.)	Tourmaline, Amethyst
Aries (23 Mar.–22 Apr.)	Bloodstone, Coral
Taurus (23 Apr.–22 May)	Sapphire, Emerald, Topaz
Gemini (23 May–22 June)	Agate, Green Tourmaline
Cancer (23 June–22 July)	Emerald, Moss Agate, Green Turquoise
Leo (23 July–22 Aug.)	Amber, Peridot, Onyx
Virgo (23 Aug.–22 Sep.)	Cornelian, Pink Jasper, Rose Quartz
Libra (23 Sep.–22 Oct.)	Opal, Tourmaline, Chrysolite
Scorpio (23 Oct.–22 Nov.)	Ruby, Topaz, Garnet
Sagittarius (23 Nov.–22 Dec.)	Turquoise, Malachite, Amethyst

Apart from the chart above which suggests the stone most appropriate for each astrological sign, how on earth does one choose the best stone for oneself from among the bewildering array of gems available to us?

Using your intuition to choose

The first and most important method is to pick what you like instinctively. The language of jewels and gems is the language of feeling. Which stone do you feel like wearing or touching? What colour, shape or texture stimulates some kind of response in you so that you instinctively want to pick it up?

Remember too that a gem may have an 'energy' for you now that is appropriate for what is occurring in your life today. By next year this could have changed. You may be drawn to a green stone today because your system needs peace and balance. Next week it could benefit from the revitalizing power of red.

When a stone has fulfilled its purpose it is important to acknowledge this. Say, 'thank you' to it. Be willing to lend or pass it on to someone else who has need of that particular quality. Sometimes a jewel will fade or crack when its work is finished. It may then be more appropriate to let it rest by displaying it in the house or even return it to the earth. This can be done by burying it in your garden, or in a window-box.

I often recharge gems that have worked hard by putting them on a crystal cluster for a few days. Trust your intuition for what is the right thing to do.

Another way of choosing is to try and sense the gem's energy with your hands. Hold your hands, palms flat, a little above different gems – one at a time – and see if you can feel a difference in vibration. It often helps to shake your hands first. By doing this you recharge the natural electrical current in them. The more you practise the more sensitive you can become to the different attributes of each stone. However, it is quite normal not to feel anything initially!

Above all, you must trust your own instinctive response. When a gem does something for you, stimulates a recognition in you, gives you what I call 'a buzz', then keep it. This stone will be of far more value than if you let your intellect decide simply from what someone else has written. It has spoken to you, opened itself to you, made you recognize its usefulness, and you can develop a mutual rapport.

Colours and healing

Each colour is a manifestation of the vibratory rate of a particular colour ray which I shall describe below. Each ray has certain characteristics. Awareness of these characteristics can be another guide when choosing gems.

The power of colour healing used to be widely understood in Ancient Egypt and by the American Indians. Its importance is being rediscovered today. Research in America showed that when violent prisoners were placed in a pink room they calmed down within ten minutes. A friend of mine had a baby that was so hyperactive from birth she thought the whole family would break down from the sleepless nights. In desperation they changed the colour of the room from bright reds, blues and yellow to a pale, pinky amethyst colour. The result was almost miraculous. Although still somewhat hyper during the day, at night the baby would lie awake, gurgling, without always creating a disturbance.

Colour is vital — in both our surroundings and the clothes we wear. Look at the success of 'Colour Me Beautiful' — the organization that helps a woman discover the colours that are truly hers. The colours that enliven her, and enhance her best points rather than make her depressed or drag her down.

As a child I used to see the 'colour' of music and hear the 'sound' of colour. This is known as synaesthesia, which is the subjective sensation of a sense other than the one being stimulated. Sometimes it was so strong I turned around quickly, thinking I would find someone there who was showing coloured lamps. In fact this still happens when I'm watching a play, an opera, or ballet. Colours pour out of the performers with every movement. I recently saw *Faust* and again caught myself involuntarily looking over one shoulder to see where the colour was being projected from. Some people associate the smell of different foods with particular colours.

I see the different octaves of colour and sound and how they penetrate everything they touch. Ultra-violet and infra-red rays are used in exactly the same way for various medical treatments. Projection of colour as a curative measure is now increasingly accepted. It is because these colour waves penetrate every level of a human being that they have such a power-ful effect.

While teaching in an ARE (Edgar Cayce's Association for Research and Englightenment) summer camp in America a few years ago I put my back out. It was extremely painful and incon-venient. A fellow teacher, a singer, came to my tent and, as I lay on my stomach, she sang a certain note. I felt as if a needle went into my spine, almost like a lumbar puncture. It was excruciatingly painful. I saw a brilliant emerald green move up my spine as clearly as if I had been injected with the colour. There was a sharp click. I stood up. The pain had gone and my spine was in place. The entire process took approximately six minutes and I was completely healed. I use both colour and sound in the majority of seminars I give and have witnessed many individual healings. Colour is a force that can change a mood, stimulate the mind and heal a wound. Conscious use of colour can help us feel and be vibrantly alive so let us now look at the characteristics of each colour.

Red

Because it has the slowest rate of vibration the most dense or opaque colour of the spectrum is red. It is a strong physical colour and can stimulate energy, vitality, regeneration of cells, blood, tissue. It is a 'hot' colour and generates heat and warmth. Red is a colour of empowerment, action, expansion to reach out and initiate something new. It is a colour that can give us a 'kick in the pants' when we feel lazy. A 'red personality' is a leader, an adventurer, an initiator of ideas — a person who enjoys life and all things physical from food to sex to sport and music. However, this person can become too aggressive and tread on other people's emotional toes if out of balance.

Red stones: ruby, garnet, agate, carnelian.

Orange

Orange is a wonderful colour for balancing the emotional body as well as the digestive system. It promotes assimilation of both food and the events of one's life. It literally helps us to digest life. It integrates what we need to accept and releases what is no longer necessary for our growth — whether this be a job, a person, an attitude or a bad habit. It helps when dealing with alcohol or eating problems. Orange is particularly effective for releasing self-pity, lack of self-worth, unwillingness to forgive. To soak in an orange bath (coloured with a few drops of cake colouring) is extremely relaxing if you are feeling emotionally frazzled. The person is usually warm, sympathetic and caring (often a nurse for example) but needs to be careful not to depend on other people's approval in order to function.

Orange stones: amber, topaz, smokey quartz.

Yellow

Yellow, like orange, is a warm colour and associated with the mental body — the intellectual, rational side of the mind. It is the colour of taking in and giving out, sharing, expressing, communicating. It is the life-giving power of the sun. Yellow can help us to develop the powers of discernment and discrimination as opposed to judgement. It is a joyful colour, good for healing the inner child — freeing us to play, enjoy, have fun. A yellow person is generally a good organizer, efficient at coping with structure, such as office management or law but needs to be careful not to let the mind isolate the heart.

Yellow stones: citrine, topaz, yellow zircon.

Green

Green is known as the fulcrum colour meaning it is the midpoint between the 'hot' colours, the earthier colours, of red, orange and yellow, and the cool colours of blue, indigo and violet. It is neither hot nor cold and therefore a major healing colour. Why do you think we are surrounded by the green of nature? Green soothes, balances, calms and because it is associated with the heart level it can help us to develop unconditional love. This means loving and accepting others as they are and not as we would like them to be, including ourselves. It can alleviate over-possessiveness and the fear of loving. A green personality is usually sensitive, responsive and loving but if his security is threatened can become possessive and jealous.

Green stones: emerald, tourmaline, jade, moss agate.

Blue

Blue, a cool colour, is associated with an aspect of mind different to yellow. It is the colour of wisdom, truth, integrity — wisdom being the blend of love and will. It can help us to be willing rather than wilful. Blue inspires mental control, clarity, creativity and the acceptance of responsibility for others. Historically it is associated with the sky, with God, with heaven. It is a colour that soothes a burn or a fever, acts as an antiseptic for bites, itches, infections. It can eliminate confusion so that we can see clearly. It is associated with the throat which is the area of speaking one's own truth. It can move us away from saying, 'If only I knew what I was here to do I'd do it'. A blue personality is usually someone who is idealistic, looking from a cooler more mental perspective than the instinctive physical love of the green person.

Blue stones: sapphire, aquamarine, turquoise, lapis lazuli, blue lace agate.

Indigo

Indigo is a cross between midnight blue and navy. It is a powerful colour which is associated with the right hemisphere of the brain and it is thought to stimulate intuition and imagination. It is a cool colour and can be used to help stem the flow of blood in haemorrhages if the patient is surrounded by this colour. Indigo is a 'psychic' colour and connects with the third eye in the centre of the forehead, clearing the veil from this inner eye, otherwise known as our intuition. An indigo person is often successful when acting on hunches and intuitive flashes. For this personality it is important not to get carried away with ideas that are unrealistic.

Indigo stones: sodalite, amethyst.

Violet/purple

Violet is the colour of the ray of transformation, change and uplifting. It is a colour I have seen pouring into the earth during the past few years and the colour most easily seen when a person first begins to practise meditation or visualization. Violet and purple are octaves of the same ray and develop inspiration. Violet/purple people tend to be artists, musicians, writers, who can reach up to another dimension and pull down for us what they see. It is therefore a colour that can stimulate a high level of creativity as well as spiritual awareness. A violet/purple person would need to guard against living in a fantasy world and being spaced out. Both violet and indigo people tend to be late for appointments!

Violet/purple stones: amethyst, clear crystal, sodalite, sugalite.

The darker and more dense a colour, the more earthy or physical it is. The lighter, more pastel shades are a different, higher, octave of the same primary note of colour. For example, pink is a derivative of red but softer, gentler, and less stimulating. The lighter the colour the higher the rate of vibration, the more solid the colour the slower the vibratory rate.

In the same way the density of a stone is an indication of its qualities. The more opaque it is the closer it is to the earth and the greater the power it has to absorb and contain. On the other hand, the more brilliant and sparkling the gem the greater its ability to stimulate and liberate our spirit and intellect. The most common error people make in choosing a stone is not to go for the one which balances them – this being the one opposite to their own nature.

Matching the stone to its user

However strange this may sound let me give you an example. A teacher and friend, with whom I was working and travelling, had so inspired his students that they decided to give him a crystal as a gesture of appreciation. This man was as wise as Solomon. Because of his great stature, physical as well as mental, these students searched, with an ever increasing frenzy

Herkimer diamond

to find the biggest, the best, the most appropriate crystal with which to say 'thank you'. Finally, in desperation, they came to me for advice. I suggested it would be far more balancing for this man's powerful personality to have a stone that was tiny, light, feminine and sparkling. He already had more than enough power of his own. He needed the opposite qualities to complement him rather than to add what was already his. They did find a 'bubbly' crystal and, believe it or not, from that moment on this teacher showed the laughing, humorous and joyous side of himself.

Another friend who always appeared to be totally 'spaced out', always bought small bubbling effervescent stones. She could never settle, left her husband, dragged her children from pillar to post, started one job and left it quickly to start another. During her travels she was suddenly drawn to three enormous crystals. They were dirty, with an orangey-earthy tinge, fairly opaque and not very beautiful, although they were tall. As soon as she installed them in her home she began to calm down. She organized regular school for her children, established a rapport with her neighbours and found a job in design that allowed her fantasy side to soar. Suddenly all the inspired airy-fairy side of her was anchored in a way that brought stability, creativity and fulfilment.

If it is our nature to be physically strong and secure we need the bubbling effervescence of a Herkimer diamond (crystals grown in water, and therefore not connected to the earth). If we tend to live in a fantasy world of mind and imagination our bodies will respond to the stabilizing, physically grounding stones of the earth, such as agate. If we are ruled by emotion we do not need a gem to increase emotional sensitivity but rather one that will enable us to act rather than react. The opposite of what we are brings balance and healing, whether in a stone or a relationship.

With that general truth in mind let us next look at some of the basic characteristics of some of the best known stones.

The healing properties of stones

Agates

Agate balances, stabilizes and protects. Medicinally the agate stone has a long history and it was prized as an amulet. Blue lace agate is lighter, calming the mind and healing fiery emotions. All agates assist physical and emotional security and can be used with other stones to balance their effect. Moss agates cleanse the immune system. They are wonderful stones to alleviate stress especially when held. They can be used almost like 'worry beads' which are popular in Greece, Egypt and Middle Eastern countries for relieving anxiety and tension. They have long been used to protect the wearer from physical injury. Any agate will assist physical and emotional security — agates can ground sexual energy and regulate the sex drive.

Amber

Amber absorbs negativity, balances the yin and yang or male and female, positive and negative aspects. Amber brings life-giving energy and the warmth of the sun. Think of throwing a stone or hurling abuse at the sun — it still shines on. Nothing can hurt it or harm it, and amber has the same power for us. It was one of the first substances used by men for amulets, medicine and decoration. You can rub it over your skin, or wear it. Water in which amber has been steeped is also a good laxative. Amber is solidified resin and when rested on the solar plexus and navel area it will help dissolve emotional rigidity — the wall-like protection that sometimes develops when a person is over-sensitive and vulnerable — like honey it has a disinfectant, antiseptic effect and I found it wonderful for post-operative scar tissue. An amber necklace can give great protection if living or working in a negative or back-biting environment.

Amethyst

Amethyst transmits, purifies and lifts. It is a powerful healing gem and stimulates intuition and spiritual awakening while calming passion, emotional violence and anger. It is also an aid to chastity! The Romans believed the amethyst prevented drunkenness and used to drink out of amethyst encrusted goblets. Its effect would be the same if amethyst was placed in a glass of wine. If you tend to be emotionally overreactive this gem will help you to be more controlled. An amethyst under the pillow will aid sleeplessness and can be put or placed on the centre of the forehead as a treatment for headaches. The centre of the forehead is the Third Eye chakra and an amethyst placed on this chakra will activate the inner seeing. It is considered to be a spiritual stone containing, in its colour, the passion, truth and love of Christ. This is a good stone to start off your investigation of crystals if you find the effect of the quartz crystal too strong. Amethyst has a gentler effect.

Aquamarine

Aquamarine is known as the water stone — the word aquamarine means 'water of the sea'. It soothes, calms and brings tranquillity and peace — washing away anxiety and negative thoughts. A friend of mind living in Canada used this gem to clear pollution from the lakes around her home. It is a 'cool' stone that will soothe a fever and relax an overactive mind. Aquamarine is a stone of protection against danger and seasickness when travelling by sea. It is supposed to bring good luck to fishermen and it cleanses the glands and nervous system and eliminates excess fluids. Drink water empowered with aquamarine by soaking the stone overnight in water. Like all blue stones it has the effect of clearing away the cobwebs in the mind. Blue stones are associated with the throat, like the turquoise. New Age thinkers believe that we are all now making a shift of identity and aquamarine will make this transitional period easier for us. It will help you to speak out the truth as you perceive it.

Bloodstone

Although bloodstone is a gem that improves the circulation and is good for all blood-related disorders (it was used to staunch haemorrhages), I have friends with hip and joint problems who feel they move more freely when wearing or carrying a bloodstone. Like the agate, it stimulates physical strength, courage and balance. It balances out deficiencies of iron in the bloodstream, cures nosebleeds, revitalizes brain tissue and helps move oxygen through the circulation. It is also a very reassuring stone to hold. Bloodstone is supposed to have formed when drops of Christ's blood fell to earth and turned to stone. Women will find bloodstone useful for regulating the menstrual flow if they are troubled by bad periods. It is a very calming stone if held in the hand.

Carnelian

Carnelian varies in colour from an orangey red to a reddish brown and sometimes salmon pink. A motivating stone, it brings, joy, happiness and blessings. The carnelian was one of the most popular gems to be used in amulets. It can help us bring our ideas to fruition and can heal jealousy and envy and can protect you from an envious person. It can energize us, stimulate courage, self-confidence, ambition, and mobilize us into action, helping people to speak out and assert themselves. Problems with digestion and appetite, including anorexia, will all benefit from carnelian. It is a stone to be used on the spleen chakra just below the navel. As well as helping in the assimilation of food, it gives us the courage to face the situations in life we need to face, to hold on to what we need to keep and to release what needs to be released. The carnelian is good for the blood, for infections and for balancing the thyroid. Mohammed and Napoleon both wore one.

Citrine

A member of the quartz family, which has similar powers to the topaz. It is connected to the sun and stimulates clarity in decision-making and also communicative abilities, receptivity, organization skills and intellectual prowess. A good stone to hold when making a choice or assessment. It is an ideal choice for the office or boardroom when communication needs to be clear and concise with ideas freely expressed. Place it on a table during meetings to help you improve your communication skills. When I'm writing I often place a citrine in my pocket or on the table near me. It stimulates the outer expression of inner ideas and creativity. Citrine opens up a bridge between the lower and higher mind, intuition and logic. It helps eliminate toxins physically and emotionally.

Coral

This simple stone from the sea is a good support for physical work. It is also protective when travelling, especially when moving from one country to another. It was originally supposed to staunch the flow of blood, cure madness and bring wisdom. Coral comes in varied shades from white to pale pink to orange to red to brown. It is another stone that aids digestion and assimilation of food and used to be used by the Romans for gum disease and toothache as well as protection against negativity. It is particularly helpful for people who worry too much about what others think of them. It can help develop this type of attitude: 'What others think of me is none of my business.' It is a balancing stone for people who work in the caring professions, such as nursing or social work. It is an excellent gift for a child to wear for their first stone as it will help to prevent them from falling without impairing their spirit of adventure!

Crystal (quartz)

This is the stimulating fire that shatters stagnation in your life, your thoughts and attitudes — it breaks open and calls into question patterns of reflex response. Crystals awaken us to higher levels of awareness. They are physically good for the heart and spine. The milky or opaque quartz is the receptive feminine aspect and the clear quartz is the stimulating male force. The person who is badly affected by criticism and turns in on themselves, broods and stews about real or imaginary slights and injustices will benefit from clear quartz. The crystal is a generator and activator of energy and repels negativity on any level. Crystallized water — water in which a crystal has soaked overnight or even for 20 minutes — will cleanse the body of toxicity and recharge the electro-magnetic field around the body.

Clear quartz can help with ear problems and hearing difficulties. Rose quartz is feminine, gentle, soothing. It is good for fears, phobias, aches, pains and general vulnerability. It prevents wrinkles and stimulates a beautiful complexion — rose quartz can heal a broken heart and feelings of inadequacy. Smoky quartz varies from shadow grey to almost jet black. It stabilizes and helps us to face our own shadow, or negativity. As you work through the problems, the shadows that prevent you from seeing the light, frequently the smoky quartz will become clear. It absorbs negativity and is comforting to hold. Rutilated quartz — with little threads of silver or gold running through it — can help focus and direct the mind. It also holds within it the connecting links to friends, family, future possibilities and past events. For more about quartz crystal generally, see Chapters 5–8.

Diamond

This is the stone of the mind because of its hard cutting edge. It symbolizes durability, incorruptibility, invincibility. *Amante de Dio* (Italian for diamond) means 'lover of God'. It is sometimes used to prevent dreaming and stimulates clear mental sight or far-seeing. The brilliance of diamonds was supposed to shine a light that kept negativity at bay. While the crystal is the gem of this age, the diamond will be the jewel of the future. In Greek mythology, Eros, the god of love, is said to have used arrows tipped with diamonds to stimulate people to fall in love with one another. It is said that a man buys three diamonds in his life: one for his wife when he marries, the second for his mistress and the third for his wife when she finds out about the mistress!

Emerald

The jewel of Venus, of love. It aids fertility, growth, honesty and self-discovery. It solves complex problems by stimulating the brain and the memory. It was once used to cure chronic infections such as the plague, by rubbing a sore with the stone. It develops clear sight, in a physical sense, and clairvoyance regarding the future. While promoting love, it can reveal the truth of a lover's promises, breaking if they are false. It is also good for bones and teeth. Emerald is used as a cure-all. It was very popular with the Ancient Egyptians and is a very good stone to wear if you are feeling vulnerable in a relationship. It heals inflammation and in the past was used as an antiseptic. I have a large emerald-green stone (not a real emerald) set in a silver ring. When my body needs peace and relaxation I hold this ring to the light and stare at it for a few minutes until I have saturated my mind with the colour. I then sit back, put the ring on my lap and imagine the colour flowing to every cell, organ, muscle and tissue in my body. After ten minutes I feel completely renewed and revitalized. This exercise can be done at any time, including when going to bed. An emerald-green stone will help the heart chakra to develop unconditional love.

Fluorite

Fluorite is a unique stone that can develop spiritual awareness. As such it is very much a stone of the future as it helps to bring down into the material reality ideas of the spirit. It is usually tricoloured − purple, white and green − and symbolizes the qualities of unconditional love, knowledge of and responsibility for the dissemination of universal love, and the power of divine will. It thus represents the threefold aspect of God. It is composed of calcium and fluorine and is good for bones (which are the structure of the body and of life), lower back pain as well as teeth and gums — and is effective in the treatment of gum disease. As a result of my accident I have had many jaw problems. I used my own fluorite to ease this area by lying down and pointing the fluorite towards my jaw. It has helped to release tightness, clicking jaw and grinding. It can also help synthesize ideas and bring people together. Fluorite is a transmuting agent that helps to ground excess energy while assisting you to operate at maximum efficiency.

Garnet

The name garnet comes from the Latin *granatus* meaning 'like seeds'. They are good for toxicity, mental depression and inactive thyroid. They balance the root or base chakra — they can stimulate repressed sexuality or modify overactive sexuality, are good for male—female sexual problems, and also for creative energy. They are stones that will encourage you to be positive and make a success out of your life. If you are a 'shrinking violet' a garnet will help to move you out of your shell. They energize the circulation and bring it to normal if deficient, such as in the case of low blood pressure. Statues of Isis often included garnets in her ceremonial belt, because they represented her blood and her power. Garnets and carnelians both benefit from being shaped or cut.

Jade

Jade comes in many shades of green and is an absorptive stone. It soothes, heals and balances; is good for heart problems and asthma and like coral is a wonderfully protective first gemstone for a child. It is a soft, gentle stone and is believed to promote a long and prosperous life. It is good for flexibility and tolerance — a stone of friendship. A lucky stone, much revered in China, it can absorb the judgement, criticism or fear that holds you back from expressing love. Taken in a drink, jade is meant to strengthen the muscles, harden the bones, calm the mind and purify the blood.

Kunzite

Kunzite, a feminine stone, is good for the skin, rejuvenates tissues and activates the heart. It stimulates self-love and acceptance, together with the ability to surrender to one's feelings. It is relatively 'new', having only been discovered in 1902, and is a beautiful pink orchid coloured gem. If a woman has difficulty in expressing her femininity then kunzite can help her accept her own femaleness. It can regulate menstrual problems which are often caused by not totally accepting one's own femaleness.

Lapis lazuli

Lapis lazuli strengthens the body during spiritual awakening, opens the mind to the divine, protects against psychic attack and depression. It enhances wisdom, inner vision and mental clarity. It is the gem of truth, integrity — the stone of the gods — and has a smattering of gold on a deep, royal blue background. It is a royal stone and was revered both in Ancient Egypt and in Jewish history, as it was probably the 'sapphire' on the high priest's vestment. It was also powdered for cosmetics used around the eyes.

Malachite

This brilliant green mineral is good for the teeth. It also soothes and calms, thus helping to foster patience, physical endurance and inner peace. It used to be considered a talisman to protect against falling and is particularly suitable for children. The Egyptians used it for eyeshadow and to promote good eyesight. It was also a treatment for cholera. It is a stone to protect and strengthen the heart chakra. It helps to clear the mind of illusion and works on all depressive types of disease. It is a fine balancer for the physical body and soothes swelling and inflammation. It has an antiseptic quality and is good for rheumatism and arthritis; is good for the pituitary gland and for men who need to be more open emotionally. Malachite may break if danger or disaster is imminent.

Moonstone

With its watery white, pearly sheen, moonstone helps develop emotional sensitivity and is supposed to arouse passion when exchanged between lovers, and to open the heart to giving and receiving love. A moonstone will balance oversensitivity and generate the ability to nurture and care for others. It is associated with the goddess Diana and legends say that its brightness increases as the new moon becomes full. It is good for the pituitary gland and can stimulate the awakening of the inner woman in a man.

Opal

A gem of psychic awareness — and of the moon and water — it can increase everything we feel as well as helping the assimilation of emotion. It is like a full moon that draws hidden emotions to the surface, causing great discomfort, and then flings it back in our face (think of the word *lunatic*, derived from the French word *lune*=moon). Opals were believed to help women giving birth to let go and relax. In the middle ages blonde maidens valued opal necklaces as they believed they helped their hair to keep its colour. Opals help the body assimilate protein. Some stories suggest that the opal is an unlucky stone. This belief originated in the nineteenth century through a story by Walter Scott. This is usually because it breaks easily and needs to be kept separately from the rest of your jewellery. Because it is a water stone, the opal likes being bathed in cold water. The fire opal stimulates passion and can heal the genital area. Opals vary in colour from translucent cream to fiery black and blue. They usually reflect rainbow colours and were sometimes used to aid prophecy and divination. They are good for the lungs.

Pearl

The pearl is very feminine, sympathetic and is also associated with the moon and water. It is emotion, chastity, purity, cures irritability, restores harmony. A pearl that loses its lustre is considered unlucky — it has lost its life-force. The pearl fishers of Borneo preserve every ninth pearl they find and put them in a bottle, with two grains of rice for each pearl, so that they will breed more pearls (so they believe). Each bottle requires the finger of a dead man as a stopper to ensure its efficacy! Because the pearl grows as the direct result of irritation it holds the key to how we can transmute and overcome oppression and pain. It is an example of how to turn the insignificant into the sublime and beautiful.

Peridot

Peridot, or chrysolite as it sometimes known, is similar in energy to the topaz. Yellowy-green in colour, peridots are brilliant and sparkling — containing sun and life — and are antidotes to depression and jealousy. They are good for the mind and liver. Peridot was used in Atlantis, Early Egypt and in the Aztec and Inca civilizations where it was used to purify the mind and the body. It aids in the absorption of digested foods and will soothe inflamed or ulcerous conditions of the intestines and bowels. Like other stones which aid the digestive system it also helps to clear emotional congestion. It is especially beneficial for the spleen chakra just below the navel and will soothe the pancreas and gall bladder. It is particularly good for the liver as it is cleansing. To drink water in which a peridot has been soaked overnight is a good treatment for the liver.

Ruby

Known as the king of precious stones, it contains within it the bloodline of humanity and the beauty of the soul. It stimulates the will to live, the initiation of ideas, the power of authority. The ruby balances mind, body, feeling and spirit. It was believed that an inextinguishable flame burned in this jewel, so strong that it could actually make water boil. It symbolizes lasting love, marriage, royalty as well as the power of the spirit. It stimulates abundance and can energize any part of the body. The ruby is connected to the seed atom contained within the human heart. This seed atom is our connection to God and contains within it the memory and knowledge of our past, present and future. The ruby is therefore a powerful stone of the spirit as well as of energizing the heart and it can activate the memory of our own Akashic record (the mind and memory of God which holds the memory of everything that has ever happened). It is much better to learn to read your own Akasha — through meditation, stillness when listening — because you are the only person who 'really knows' your own history.

A psychic or clairvoyant can attune (like dialling to a particular radio station) to the idea before an event, to the event itself, as well as to the result. A psychic does not necessarily tune into all three at once; you could therefore go to three psychics and get totally conflicting results. They may be part of the same story but different, depending on what actually took place and the psychic's interpretations which also depend on their own level of development. I could, for instance, decide to paint my house purple — which is the idea before the event. I start to do it but see that it will be too strong so halfway through I change it: the result is that my house is cream. Now if you had only tuned in to different portions of the event you would say that you will and are painting your house purple — and then someone else may say that you have a cream house. In that situation it doesn't matter but if it's dealing with a life-changing situation it does — you are the only person to know what is best for you. By holding and attuning to a ruby, you can learn to get in touch with your Akashic record and can also learn to listen to your heart.

Sapphire

The sapphire's deep blue shades are due to the presence of ferric iron or titanium, while other colours of sapphire, such as pink, green or yellow, come from other mineral compounds. It is a jewel of truth, beauty, clarity and wisdom. It is an astringent for bathing and healing infection. It is also associated with heavenly virtue, devotion, self-control – and is particularly useful for controlling desire. The Ten Commandments were engraved on stones of sapphire – although it is thought that this may in fact have been lapis lazuli. The sapphire, especially the star sapphire, is the stone of destiny that can stimulate us to reach for the stars. Some cultures believe that the three light bars which cross to form the six points on the surface of the gem represent faith, hope and destiny. If you have had an idea or an ideal, especially if it would help humanity, a star sapphire can help you hold the dream and make it a reality.

Sodalite

Sodalite is a deep purple-violet blue. It contains within it the energy to increase spiritual awareness. If held in the centre of the forehead, against the Third Eye, it can stimulate the opening to a new perspective. It is a stone for the 1990s, creating a link between conscious, superconscious and subconscious. It balances the metabolism, lowers blood pressure and helps sleep. If one is drawn to this stone, it is often because of physical deficiencies within the body and sodalite can balance the metabolism, especially the body's production of insulin. The power of sodalite is amplified when used with another gem such as clear quartz.

Sugalite

This pink, violet, blue stone is very beneficial to the development of the Third Eye seeing or inner vision. It will clear away mental cobwebs between intuition or right-brain thinking and logic or left-brain thinking. It balances the pineal and

pituitary glands and can help children with learning difficulties, and also with autism, dyslexia and epilepsy. It is
another gem of 'now' and also a wonderful aid to meditation.

Tiger's eye

This is a rich, earthy stone which steadies, solidifies, brings
together and condenses wild thoughts and imaginings. It can
help us to concentrate and to bring our physical energy into
sharp focus to accomplish a particular goal. Tiger's eye, sometimes called 'cat's eye', is meant to help you see in the dark,
develop insight and bring luck. It works on mass consciousness,
helping us to separate false desire from real need. It also helps
to purify the body after over-indulgence in rich food. A fairly
dense though lustrous stone, in the process of 'becoming' or
evolving, it can help us move from density to light while giving
protection in the process.

Topaz

It is believed to prevent colds and tuberculosis because it can
activate the lungs, and has similar qualities to amber. Topaz
means 'fire' in Hindu. It brings light to life, relieves stress and
protects against danger. It comes in yellow, brown and blue,
the gold and amber shades being the most valuable. If you put
a topaz under your pillow at night it will soothe the nerves and
energize the body during sleep.

Tourmaline

The name comes from the Singalese name 'tourmali' which
means a mixed-colour precious stone. The most popular type is
the watermelon pink and green combination. Because of its
ability to produce negative and positive charges at either end
tourmaline is a wonderful balance. It can stimulate communication and co-operation between opposing forces. It is tranquil,
calming, and good for lymphatic and digestive systems. One of

the most versatile gems, it affects mind and body and does not absorb or hold any negativity. Watermelon tourmaline acts upon the wearer as a polarity balance and redirects energies that may be in the process of being congested or obstructed. The current transition on the Earth is balancing the male and female and by its own balancing effect this gemstone eliminates misunderstanding, bigotry and intolerance, and is therefore another stone of particular value for the 1990s. It can assist in the forgiveness of self and others and is particularly helpful when one is sensitive to criticism and other's opinion. It allows the current to flow between mind/ spirit and body/feeling. By its combination of pink and green (pink within) it symbolizes the tunnel experienced in death or NDE (Near Death Experience) and can facilitate the movement of consciousness from one level to another — death and rebirth — remember that birth here is death on a different dimension, and death here is birth on another level. This stone is full of joy and can help to heal a broken heart.

Turquoise

To the Atlanteans and American Indians the turquoise symbolized the sky, the breath of life and spirit, and was a reminder that man is a creature of spirit rather than flesh. In the fourteenth century it was said that the turquoise 'protected horses from the ill effects of drinking cold water when over-heated by exertion'. It is a stone of courage, fulfilment and success. It is a marvellous stone (or colour) to wear around the neck if one is afraid of public speaking. The turquoise represents strength and power. It is sacred to the Pueblo Indians of New Mexico and until the end of the seventeenth century it used to be worn only by men. It is a soft stone and can sometimes lose its colour. It is particularly good for the throat, for the lungs and respiration, and, because of its high copper content, it is a wonderful conductor of healing. It is also good for rheumatism and arthritis.

Crystals and how to choose them

'Crystals sing and speak. When they want to come to you they sparkle and shine. They say, "Look at me – here I am. Take me . . . me . . . me". They pop into your life in mysterious and magical ways. They are a law unto themselves and when they choose you, if you respond, you will feel an inner tug that is almost irresistible. If you pick the crystal up you will find yourself bonding with it and it will be very hard not to take it home. It can be the beginning of a relationship that will make your life happier and healthier.'

A man and a woman in the front row of the audience had had enough. They stood up and, directing a look of disgust at the speaker, and amazement towards the apparently rapt audience, moved to the door.

As they went out they passed a sign, 'CRYSTALS – PATH-WAY TO THE STARS. EVERYONE WELCOME', which had led them astray, after a hard day's work, from their normal route home.

A woman in a red dress near the back raised her hand and said, 'You don't honestly expect us to believe that you can bond with a stone and live happily ever after, do you?' Amidst general laughter, and after she had given an explanation of her format for the following day, the speaker brought the evening to a close. Most of the people present had a sense of excitement and anticipation about what would unfold during the next two days. A few, experiencing similar feelings to the couple who had left, obviously thought that anyone who talked about stones speaking was completely crazy.

I shall always remember this particular seminar. For a start I was the speaker. Secondly, it was in the early days of crystal workshops and few people in the audience had ever thought more about the mineral kingdom than when they had chosen a gem for an engagement ring. Thirdly, wonderful and miraculous things began to happen for all of us who were there.

The next day I arrived early, in order to arrange dozens of crystals in varying shapes and sizes on black velvet-covered tables. On one table I placed a group of crystals — almost like a crystal bouquet — comprising some of my biggest and most unusual stones.

There were crystals for sale, crystals to experiment with. There were crystals galore to touch, hold, feel, be dazzled by. There were big crystals, small ones — some bright-shining clear, others more opaque. The whole room began to glow from the glittering, shimmering mass of stones displayed.

As people began to arrive I was surprised to see the woman who had made the adverse comment the previous evening come to the table and pick up a crystal. After she had examined it closely for a minute or two she replaced it and went to her seat. Almost immediately someone else picked up the same crystal and did *not* put it back. She carried it away, sat down and kept it on her lap. The first woman became increasingly agitated and after ten minutes finally came to me and said to me in anguished tones, pointing a finger, 'That person has taken *my* crystal.'

Before I could reply I felt a tap on my shoulder. Turning, I was even more amazed to find, standing behind me, the couple who had left in such disgust the day before. Not only that, they were also clutching a very clear and shiny crystal, about the size of an index finger. The husband, whose name was Tom, could barely get his words out fast enough. 'We found this crystal sitting in a puddle of mud on the way home. It just looked at us. It was waiting there — how could it have got there? — it's just not possible. We had to pick it up and take it home ... Yesterday we thought you were all bonkers — today we decided we had to come back and find out more about crystals. The crystal made us come ...' His voice tailed away. We all laughed.

In each case it was the crystals themselves which had worked

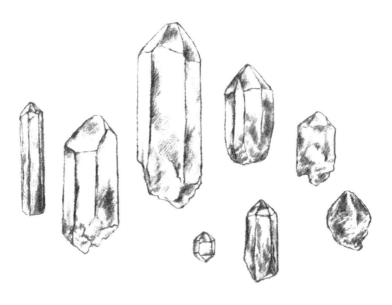

Quartz crystals come in many different shapes

some kind of magic that opened these three people to an awareness of something beyond the rational mind. Tom, and his wife Jodie stayed for the entire course. They became crystal fanatics and two years later decided to open a crystal shop. The other lady, Maureen, who was so upset when somebody else took *her* crystal, was able to retrieve it, buy it and live happily ever after with it. In fact, putting the crystal under her pillow helped her to dream. Working with her dreams gave her so much insight into her life that she became a radically different person. This in turn brought about a change in her relationship with her husband, children, friends and neighbours. From the moment the crystal introduced itself into her life everything blossomed.

I share these stories because crystals are far more than just their geological make-up, their number of facets, their power 'to oscillate' at specific vibratory rates in radios, watches and computers. We really know only 'so much' about them. Setting aside laboratory-grown crystals, we do not even know exactly how long they take to grow. There is an X-factor about them that even scientists have not been able to decipher.

We do know that crystals are activated by heat, sound and pressure along their axes. When energy moves along the axis it is amplified. This means that the energy field around the crystal expands (5 cu. ft per lb of crystal) and creates an inner pulsation which you can feel if you hold it. However, crystals are also activated by, and respond to, the energy of thought and feeling, especially love. Maybe this indecipherable X-factor is love.

How wonderful to think that when a crystal singles you or me out, individually − stimulates a response that is *not* experienced by a husband, wife or best friend − it may in fact be saying, 'I love you. I care about you. I want to be with you.'

Therefore for me the most vital factor in choosing a crystal for my own personal use is how I feel about it. How do I know how I feel, if, faced with a display of two hundred crystals, not one of them seems to be saying, 'take me'? First I look − without touching. After a good look, if a crystal attracts my eye, I will pick it up and see what happens, how it affects me. Second, I run my hands over the crystals, holding my palms flat two or three inches above them. I do this with my eyes open or closed. Sometimes I feel a strong pulsation. This procedure is what you must follow too.

If you are choosing a crystal for the first time, try not to let the logical, left-brain part of your mind decide for you. Crystals speak through feeling and intuition − heart feeling combined with physical sensation. By trusting your feelings and intuition, when you hold a crystal, you are speaking the same language.

It is advisable to pick the crystal up in your right hand − unless you are left-handed − and hold it near the centre of your chest around the level of your heart. You hold it for a few minutes, rotating it so that each facet, or side of the crystal, has been felt near to, but not touching, the chest. (See illustration on page 80.)

You may get a sensation of coolness and warmth. Your heart may give a slight flutter or quicken its beat. This certainly happens to me when I sense a rapport with a crystal for the first time. The hairs on your arms may stand on end. You may experience the same sort of shivery feeling we get when we say 'Someone's walking over my grave.' You may feel none of these things but simply know from a deep inner part of yourself that this is the crystal for you. Trust that feeling.

Choosing a crystal by holding it near the heart

Once you have established some kind of connection, try the crystal in the other hand. Hold it with the sharp-pointed end directed up your arm, then pointing down. Move it over your legs, over your body. Do this with both left and right hands. Listen to it with your feeling, your imagination, your intuition.

If, having done all this, you still experience no real response or sense of relating, then put the crystal down and try another one. You may have to do this many times, even going to another shop if necessary, before you find 'your' crystal.

Whatever you do, do *not* hand a crystal to someone else and buy it on the grounds of what he or she says and feels. Crystal energy is highly individual. Just as the personalities of the people around us affect us all in different ways, so do the qualities of crystals evoke a different response in different people.

This is the way I choose crystals if I am looking for a joyous and loving companion. On the other hand when I need a crystal for a specific purpose, such as directing healing energy into a broken bone, or drawing pain out of a burned hand, I will look for the type of crystal that is most suitable. I use logic to find the crystal that will best suit the purpose. It may need to be opaque or clear, male or female, a pusher or puller, left-handed or right-handed. Each crystal, by its shape and clarity, lends itself more easily to one task than another.

The structure of crystals

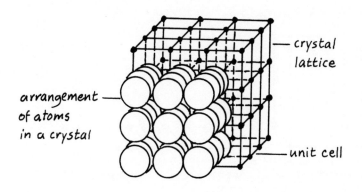

The arrangement of atoms in a crystal

Crystals and gems are characterized by regular arrangements of atoms. Non-crystalline matter has its atoms scattered around and not specially arranged. Six of the basic shapes of the building blocks of crystals are as follows.

1   isometric – a straight geometric cube, e.g. fluorite
2   tetragonal – vertical faces longer than horizontal, e.g. wolfenite
3   hexagonal – a six-sided prism, e.g. emerald
4   orthorhombic – lozenge shaped, e.g. topaz
5   monoclinic – has three mutually unequal axes of which two are at an oblique angle, e.g. azurite
6   triclinic – has three mutually unequal axes all set at oblique angles, e.g. turquoise

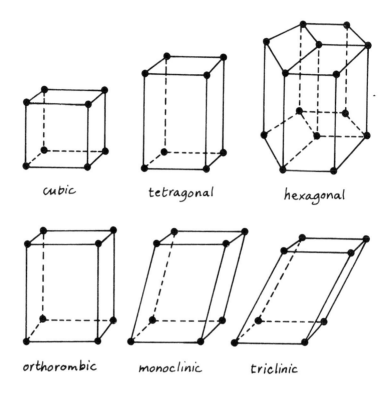

cubic   tetragonal   hexagonal

orthorombic   monoclinic   triclinic

The six basic building blocks of crystal

These are the only shapes which can be used to build a solid three-dimensional object. Bricks used in the walls of houses are the same shape as the orthorhombic building blocks. If any other shapes were used the final result would have holes in it.

Crystal facets and numerology

As my relationship with crystals developed I discovered that I was frequently attracted to a stone with a specific number of facets that had significance to my life. I based my interpretation of the numbers of crystal facets on numerology. Numerology is the interpretation of the symbolic qualities of numbers. Lynn Buess's book *Numerology for the New Age* says that the science of numbers dates to pre-Atlantean times and it was used in both Hindu and Arabic teachings. Pythagoras believed that divine laws are reflected in the mathematics of numbers. The *Kabbal*, the book that enshrines the occult and esoteric tradition hidden within the Jewish religion, goes deeply into the essence of numbers. In the Chinese tradition of numerology the odd numbers are thought to be yang, or masculine. These odd numbers are believed to bring good fortune and to represent the celestial world. Even numbers are yin or feminine, and represent receptivity and the terrestrial world.

In the language of numerology **one** is the number of beginning, of initiating new projects, new beginnings. It is a number of inspiration, ideas and confidence.

Two is the number of duality, of reaching out, caring and sharing. It is the symbol of mothering. The two coins mentioned in the parable of the Good Samaritan symbolize this ability to nurture and protect. Two crystals joined represent partnership and sharing.

Three is the number of communication, expression and creativity. Three represents the threefold aspect of God, trinity of Mother, Father and Son as Wisdom, Will and Love. It is the triple power of birth, life and death. Some crystals have a small triangle etched on to one facet. I always feel that these crystals contain a special and divine message. They have received a unique stamp from the gods.

Four is the number of mastering the laws of the Earth and it usually involves hard work. Four is frequently used as the number of wholeness, the four quadrants that equate balance of mind, body, spirit and feeling. Many crystals have an extra four-sided facet, often diamond shaped, on the main termination face. If you meditate or concentrate on this you can get insight about the major task of your life and how to accomplish it.

Five is the number of freedom, versatility, curiosity and expansion. It is the number of the perfected man with his feet on the earth and his arms reaching for the stars. A five-faceted stone, although not so common, will help you move beyond the restriction of other people's ideas and influence.

Most crystals have **six** facets around the termination point. Six is the number of balance, harmony and synthesis. A six-faceted crystal will help you bring into balance your relationship with yourself and therefore your relationship with others. It will open your heart to give unconditional love to everyone around you.

Seven is the number of the mystic, the seeker of truth, who looks at life from a different perspective. There are many references to seven in the Bible such as Elisha raising the child from the dead seven times, the dove sent out from the Ark on the seventh day. There are seven days of the week, seven branches on the tree of life and seven branches on the Jewish Minorah. A seven-faceted crystal will enable you to discover spiritual truths and penetrate the mysteries of life.

Eight is the number of power. The two circles, one above the other, symbolize the macrocosm and microcosm, as above so below. An eight-faceted crystal will stimulate the power to assert and express yourself, to 'be who you are' without holding yourself back. It will also help you to cope better with power, money, possessions or position.

Nine is the number of mastery over the lessons of life. If life is a school, a number nine crystal will enable you to pass your final exams, overcome the challenges of illness, loss of job, change on any level.

Ten is the number of 'I am whole'. It combines the one of 'I am' with the circle of completion. A ten-faceted crystal develops a sense of oneness with God.

Eleven is considered to be a master number, as are twenty-two and thirty-three. A master number has added potential. It is like double dose of the qualities attributed to the original number. Eleven is therefore two 1s and emphasizes the authority of the number 1. 1 plus 1 equals 2 and so eleven also symbolizes the mastering of the ability to care for others — to accept responsibility, to reach out from a position of authority to teach or organize.

Twelve adds up to three and completes a cycle. There are twelve signs of the zodiac, twelve hours of the day and night, twelve disciples. A twelve-faceted crystal symbolizes cosmic order, the trinity, the blend of world and spirit and the ability to express that blend.

In Aztec calenders **thirteen** was a number of divination. One plus three equals four and so it also symbolizes hard work but from a different, more far-reaching perspective than four on its own.

One of the first scientists to study crystals was Nicolaus Steno, a seventeenth-century geologist and Catholic priest. He formulated a law known as 'The Constancy of Interfacial Angles'. This states that in all crystals the angles between corresponding sets of external surfaces or faces are always the same. Steno's discovery hinted at underlying order in the universe. In 1784 the discovery was elaborated by Abbé Haüy, Professor of Minerology at the Museum of National History in Paris. Haüy was also a Canon at Notre Dame. Both he and Steno 'believed' they had discovered an aspect of the mind of God.

The accuracy of their conclusions was later confirmed in 1912 by Max van Laue, a German physicist, using X-ray crystallography. The way in which the atoms are arranged inside a crystal is responsible for its outward shape. By measurement of the angles between the faces of a crystal one can discover on which of the building blocks it is based. So incredibly exact is the internal atomic structure — or lattice — of these crystalline building blocks that the Egyptians used them as the base to their pyramids.

When I first began to experiment with crystals I knew nothing of crystalline building blocks or the structure of atoms. I had never heard of Steno, Haüy or von Laue, and X-ray crystallography might have been the language spoken on another planet. Dreams and meditation showed me that it was time to explore the world of crystals and so I did.

I saw the clear pointed end — or termination — of the crystal as male and the usually more opaque, opposite end as female. I sensed a very different energy coming from one point compared to the other. I sometimes saw red colour coming from the male point and blue coming from the female. A more scientific friend of mine scoffed at my ideas of male and female crystals. However I later showed him Kirlian photographs that illustrated exactly what I saw. This is a form of photograph invented by a Russian, Peter Kirlian, which is able to register the electromagnetic field surrounding people, plants, animals and stones.

In fact all minerals that are opaque, through which no light can be seen, absorb rather than transmit light or energy. Transparent minerals are those which can be seen through clearly and these are the best for transmitting and amplifying. Translucent minerals are those through which a little light can be seen.

The kinds of atoms and the arrangement of atoms inside the crystal determine whether or not light can be transmitted through it. In transparent minerals light waves striking the specimen make the atoms vibrate. The vibration is passed from atom to atom through the lattice structure. Opaque minerals have atomic arrangements which are unable to transmit light waves and the vibration is quickly absorbed. Translucent minerals allow some light to be transmitted while some is absorbed. The lustre of a mineral refers to its surface appearance and again depends upon the ways in which the atoms and their arrangements affect the incoming light waves. It was not until much later that I discovered all these scientific facts and explanations for what I sensed. Initially I worked from intuition, trial and error.

Trial and error is a great teacher. What is more, it is free, available to everyone, and there is no better way to learn about crystals. What affects me in one way may affect you in another. Try, experiment, find out what works for you.

Varieties of working crystals
Double-terminated crystals

I am frequently drawn to crystals that are DTs — or double-terminated. This means natural points at both ends (by natural I mean they are not cut or faceted; they grew like that). A single termination means a single point at one end. The other end is often uneven from where it was broken off from its foundation. A double-terminated crystal contains within it the two opposites of negative–positive, male–female, while uniting them in the middle. Some DTs, which break at one end during their formation, heal themselves by growing the second point over the wound. These are doubly effective in rebuilding the physical body because they demonstrate the power to heal oneself.

Double-terminated crystal

Because crystals respond, amplify and transmit, at a very high vibrational frequency, they have the ability to open us to other levels of awareness. They affect the energy field around us, known as the aura, which contains within it the pattern of our lives, thoughts and emotions. Crystals pulse energy into whatever is within a three foot radius and therefore help to shatter structure or crystallization – especially in the auric field. All crystals will do this, large or small. However DTs, which contain within them the power of opposites, can help us explore life—death, fear—joy, black—white and good—evil. In other words they stimulate us to explore the extremes in ourselves — which is not always comfortable — while bringing us to a point of balance in the middle. DTs symbolize the uniting of opposites. They can be useful in communication problems because, by pushing us to open closed doors, they help us to listen to the *opposition*. They are both transmitter and receiver — they are like a double dose of transformation. If this is what you need, then by all means choose a DT. If you are new to crystals, you may find the single-terminated crystal a better or more gentle choice.

Single-terminated crystals

Single-terminated crystals are six-sided and usually have six facets or faces around the termination point. I say usually because sometimes there is an extra facet — diamond or lozenge-shaped — on the left or right of the central face (see illustration). If this is on the right side as you look at it it means

 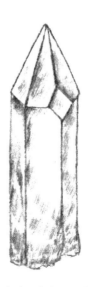

Left-handed crystal   *Right-handed crystal*

the crystal is predominantly right-handed. Without these dis-tinguishing marks most crystals are a mix of both. The right-handed crystal is more effective for the right side of the body, the left-handed one better suited for the left side of the body.

Right and left-handed crystals

Unless you are left-handed, the left side is controlled by the right side of the brain and vice-versa. The right brain is the intuitive, imaginative, receptive part, while the left brain is the logical, rational, deductive part. One feels and listens. The other thinks and acts. Therefore the left side of the body is considered to be the receptive or feminine, the right side the assertive or masculine. If you need to be more understanding, more open to feelings, a better listener − if you want to develop greater rapport with your intuition − the left-handed crystal is the better choice. Or, if you need to be more assertive, decisive, take action in some way, whether mental or physical, the right-handed crystal is more appropriate.

Transmitter crystals

A transmitter, or pusher, crystal is one that has a clearly defined three-sided pyramid face, like a triangle in the centre (see illustration). This triangle can be quite big or very small and usually has two seven-sided faces on either side. (I say usually because the crystal in front of me now has a seven-sided face on one side and a six-sided face on the other.) This type of crystal when held can help to direct or transmit healing energy. It does not replace what can be done with the hands but it amplifies it − as a microphone amplifies a voice. A transmitter can also be used for absent healing by placing it over the name of the person who is ill or by holding it while you concentrate on what that person most needs to be cured. (Do not specify a particular energy unless it is love or light!) It can be used to communicate feelings of love to family or friends who are far away, almost like a personal broadcasting system. You can ask questions of the universe, guardians or guides, through it.

A homeopath I know uses a transmitting crystal to amplify

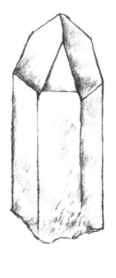 

Transmitter or pusher crystal   *Receiving or puller crystal*

her work. She mixes her potions in a crystal-filled room and at different times during the day she broadcasts healing energy to her patients via the crystals. In one case the mother of a small boy with acute earache telephoned to say that, in spite of the remedy, the child was still screaming with pain. What else could she do? The homeopath immediately directed healing through a transmitter crystal. Within ten minutes the mother phoned again to say that her son described a sensation like a laser hitting his ear, after which the pain stopped.

Receptive crystals

A receiving or receptive crystal is one with a seven-sided flat face as its main facet (see illustration). This is the type of crystal that in healing can help to draw or pull pain out. For this purpose it is used in the left hand. Recently a friend of mine accidentally dropped a blackboard on her foot and broke it. While awaiting the doctor's arrival I used a receiving crystal to draw away the pain and trauma. The doctor then encased the foot in plaster. During the next few days I worked on it five or six times a day, for about ten minutes each time. She also took the crystal to bed and kept it on the floor near her foot during the day. She experienced almost no pain. Seven days later the plaster was removed and the doctor could not believe how quickly the bone had healed. She still had to hobble but she managed without the plaster far quicker than is normal.

The receiving, receptive crystal, sometimes called a puller or a channelling crystal, is particularly useful for meditation when held in the hand or on the lap. It will store, or record, the memory of the meditation, thus making it easier to relax more quickly into a meditative state each time it is used. Receptive crystals are good listeners and stimulate the capacity to listen or tune in to our own inner wisdom and guidance. If kept in or near the bed they will help us to dream and remember our dreams.

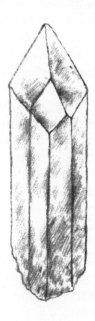

Window crystal (note the window effect where the faces join)

Rainbows in crystals

Crystals with rainbows are particularly joyous and loving. I see rainbows, whether real or in a dream, as a blessing from God, a sign of angelic protection and love. (Look at the rainbow in the amethyst picture opposite p.106). Often the colour prisms which form the rainbows in crystals are there as the result of a hard knock or a flaw. For this reason too they are joyous companions reminding us that our own hard knocks and flaws may bring us beauty and joy. If you feel depressed or disillusioned by life try keeping a rainbow crystal in your pocket. You will soon feel better. The rainbow is a bridge between this and higher worlds — the rainbow crystal helps us to bridge the gap.

During one of the most devastating periods of my life I went through months when I could not sleep or eat. I felt isolated, alone — betrayed by life and everything I believed in. It was as if thick, black smog hemmed me in on every side. I thought I would drown in it. A rainbow crystal came into my life. Almost without thinking I put it under my pillow. That night I slept and had a dream. In the dream I faced a huge mountain made of granite steps. Each step was as tall as my body. By standing on tiptoe I could just reach the next step with the tips of my fingers and I dragged myself up. By the time I reached the top my body was raw and bleeding. I looked up and found myself confronting thick, black, evil-looking slime. It had no shape — it was just everywhere. It filled me with revulsion. It was the sum total of all that had happened and all I feared could and would happen. I moved to go back down the mountain but when I saw the height I was too afraid. Turning to the slime I smashed my fist into the middle of it. It burst into thousands of crystal rainbows which cascaded around me in an explosion of light and colour. I woke up electrified. I realized that by retreating from the blackness — because of the fear that it would engulf me — it expanded and grew. As soon as I faced it, touched it, it was instantly transformed and beautiful. This dream, combined with the rainbow crystal, changed my understanding and gave me the impetus to change my life.

Herkimer diamonds

A crystal with abundant joy, sparkle and bounce is the effervescent Herkimer diamond (see page 60 for picture). This crystal is only mined in Herkimer County, New York State, USA. It is a member of the quartz family but glitters and shines like a faceted diamond. Herkimers grow in a liquid solution, in a host rock of dolomite, as opposed to a silicate type rock. They are therefore very different, both chemically and in their effect. Quartz crystal grows from an earthed foundation — it is connected to the earth. The Herkimer forms without attachment to the earth — it is suspended in the liquid, inside the host rock, and can grow in any direction without inhibition. For that reason it lifts the spirit, expands and frees us from the weight of too many responsibilities. If you need a little joy, laughter, spontaneity — a little more dazzle in your life — try a Herkimer. Despite its effervescent qualities it contains the power to free the spirit and let it soar.

Laser wands

A crystal that, combined with visualization, can be used like a pair of scissors or a knife, is the laser wand. It is long and narrow and not particularly beautiful — it is more like a surgical instrument. Mine are greyish in colour and appear almost dirty. The laser wand is like a sword of light — think of Darth Vader's sword in *Star Wars* — and can be used to cut through mental and emotional entanglements. It severs the bonds that stifle and restrict us.

A girl, Mary Jo, came to see me and immediately picked up the two or three laser wands I had with me at the time. When she understood how to use them she borrowed one, and later told me that she felt hampered by her parents' constant pulling on her emotional strings. To change this she visualized them in front of her, imagined the bond between them and mentally cut it using the Wand. She did it with an affirmation of love and peace, lifting the relationship to a new level of understanding and acceptance. She said, 'I love my parents but I'm now thirty-eight years old. I have lived my life tied to my parents' apron-strings. I now feel as if fresh air is blowing between us. I feel free to make my own decisions unhampered by their opinion.' Remember that if you are unable to find a laser wand you can programme any crystal to do the same thing.

Phantom crystals

A phantom crystal is one with an inclusion of another element. At some stage the crystal stopped growing and another material was deposited on the crystal faces. When conditions changed, the crystal started to grow again, around the trapped particles. The phantom outline shows where the growth stopped — it is like a ghostly image of another crystal. Because an element of nature has merged with the crystal, phantoms can open us to the wonders of nature and stimulate greater co-operation with the elemental kingdom. If you want to start a conversation with your plants use the phantom as a go-between! A gardener friend uses her phantom almost like Aladdin uses his lamp. When she needs help in the garden, especially when a plant is ailing, she rubs her phantom and summons Fred, a little gnome. I have never seen Fred but she swears he runs about in bright red Wellington boots! Phantoms inspire imagination and creativity — they spark us up to achieve the impossible.

Twin crystals

Twin crystals are two crystals connected to each other. They can be equal in size but sometimes one is much bigger than the other. They are mutually supportive and contain both left and right polarities. They will balance our inner masculine and feminine aspects, as well as our external male/female relationships. Penetration twins are crystals growing together in such a way that they appear to penetrate one another and tend to have a common centre. They symbolize unification of ideals, the ability to merge with another without loss of power — individuality separates, uniqueness recognizes 'I can merge and still be me.' This is a crystal for couples or someone needing to unify a group. Triplets, when three crystals are joined, are good for communication. Communication with self, with others, with intuition and inner wisdom.

Crystal clusters

Crystal clusters, when many crystals grow together from the same base, are wonderful for families, and the home. Each person does his own thing while moving out from the same home base. Clusters contain the polarities of negative and positive and, in the bigger clusters, the directions of North, South, East and West as well. They clear the air, almost like an ionizer, and recharge the atmosphere. A cluster placed in an office, a waiting-room — anywhere people congregate — creates harmony and balance. Clusters stimulate smooth efficiency in working situations. They are also wonderful in the garden.

You will only get into high numbers of facets by counting all the faces on double terminated, twin, triple or many clusters of crystals. They will add up to one number such as twenty-five which is $2+5=7$. I usually look at the qualities of each individual number in this case, as well as the final sum total. It gives a broader interpretation of why I have been drawn to this number of facets. If I feel in need of the qualities of a particular number I will count the crystal facets. Otherwise I let my intuition and feelings guide me.

When choosing crystals for bodywork — crystal healing and massage — it is more sensible to use stones of 7–12 cm (3–5 in). The power of a crystal is not dependent merely on size but a bigger crystal covers a larger area quicker. With techniques that require the crystal to be held point down, the point needs to extend at least an inch from the hand. It also needs to fit comfortably against the palm (see illustration). Do not get a

When holding a crystal it needs to have a firm connection with the palm of the hand

crystal that is so big that you cannot manage it easily. You will find that your hand gets tired and it will interfere with the healing. My own working crystals are approximately 12—15 cm (5—6 in).

How to set your crystal

Crystals for personal use that you want to carry or wear need to be small and light enough not to weigh you down. If you wear your crystal on a chain do not use a clamplike fixture which

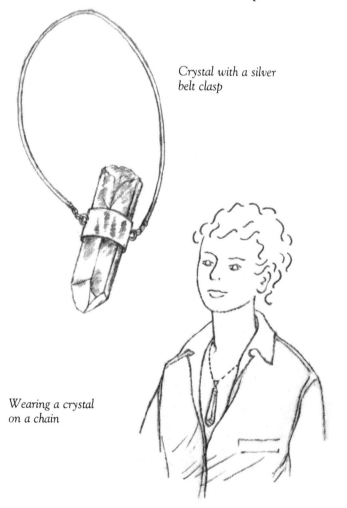

Crystal with a silver belt clasp

Wearing a crystal on a chain

closes off one end. A crystal breathes. It therefore needs a belt-like clasp around its middle which leaves it free to breathe at both ends. Silver is the best metal to use with quartz crystal.

Once, on an aeroplane, I saw a crystal which a woman was wearing on a chain around her neck. It looked breathless, set in a heavy gold clasp. Trapped next to her in my seat, I could bear it no longer. I said, 'I know this sounds a little odd, but the crystal on your chain is desperately unhappy. I suggest that you have it unclamped.' The owner was initially taken aback, but later told me that she had always felt slightly uncomfortable wearing it, breathless and restricted herself. She promised to follow my advice. She was in fact sensitive to the crystal's vibration but had not recognized it.

No matter what is written about crystal qualities only you can truly decide what is right for you. Some crystals will enable us to feel joyous and uplifted, others create a sense of expansion that may benefit dreams and meditation. There are crystals that balance us emotionally or clear away mental confusion so that we can make decisions and choices from a point of clarity and peace. I have one crystal that has the energy of a Samurai swordsman. It gives me a feeling of protection. Another anchors and stabilizes me if I find myself getting disconnected from the earth and everyday living. Some crystals make us more assertive and courageous, while others open doors to creativity and imagination. My Samurai swordsman may create in *you* a totally different feeling to the sense that *I* have of being protected. You have to read what is written and then virtually throw it out of the window and trust your own perception of how a specific crystal affects you.

Your thought and interest regarding crystals puts out a message into the universe that magnetizes them into your life. Remember the principle that 'energy follows thought'. It is an aspect of universal law. When you 'put out' a thought you also create a space. The universe hates imbalance so it fills that space with what you thought about. You 'draw to you' or magnetize what you think. This is called synchronicity. When you think about crystals you'll find they start appearing in unexpected places. Remember Tom and Jodie who found a crystal 'sitting in the mud'. If the idea of finding a crystal on the road or in the garden like that sounds too far-fetched, I can vouch for its authenticity. I personally know a number of people who, after attending a crystal workshop, or reading a book, became curious about crystals, then literally stumbled over them just like that. One woman found five crystals in her garden the day after a workshop. Maybe they were there all the time and she was simply unaware of them. Maybe they manifested themselves. Who knows? One man picked up a crystal on the beach, another man found three in the back of a drawer when tidying up; a third was given a crystal as a surprise present by his four-year-old daughter, who had discovered it herself in a pond.

When a crystal appears unexpectedly in your life it is a special gift from the universe to you — it has grown itself for you. Treasure it. Whether you buy, find, or receive crystals as presents, they are all very special gifts from the universe. Do not, therefore, get fussed, anxious, confused or too serious about choosing. The correct crystal for you will always turn up.

'Can I choose a crystal for someone else?' is a question I am often asked. Yes, of course. Think of yourself as the channel through which the crystal reaches the person it grew itself for. I suggest that while looking at a selection of crystals, you think of the person's name. Even say their name, silently in your head, visualize the way they look and eventually ask the crystals 'which one of you is for so-and-so?'. You can do this with your eyes closed or open, whichever you prefer. Then let your hand be drawn to the one that feels right. Trust your intuition and the consciousness of the crystals themselves. Be assured that if you let your inner sensing guide you, you *will* find the crystal that is correct for you, or a friend, at the moment when you most need the quality that particular crystal can provide.

Another question that crops up concerns the difference between man-made and natural crystals. Man-made crystals — glass, or lead crystal — have a high content of lead oxide which is insulating. They have no life-force and do not transmit energy. Man-made crystals are also affected by the thoughts of the men who made them. Lead crystal, in myriad shapes — suns, moons, stars — is delightful and decorative. With sunlight it can magically fill a room with rainbows. However it does not have the capacity to heal or transform, apart from in a purely visual sense.

Crystals in divination

Crystal balls

Crystal balls for scrying or crystal gazing are sometimes made from lead crystal. They are very beautiful but I personally prefer the ones made from real quartz. Crystal balls are used in divination as a focus to attract the attention of the gazer. The points of light reflected from the polished surface help to fix the eye until the optic nerve becomes fatigued and stops transmitting to the brain. This lowers the brainwave rhythm into a slower, or Alpha, state which allows the brain to produce its own images, or to see with the inner eye rather than the outer. Crystal-gazing has been practised since Egyptian times and is still very popular for fortune-telling.

Crystal used for dowsing

Dowsing with crystals

Another form of divination is dowsing with a pendulum. Dowsing used to be the term used for finding water by means of a dowsing rod. The dowser held the rod as he walked over an area of land. If there was water the rod quivered violently. Nowadays dowsing is also used to denote the use of the pendulum. A pendulum can be made from virtually anything – a stone, a twig, metal – or a crystal. It is attached to a thread or a fine chain and one can learn to use it to get yes or no answers to questions. It swings one way for yes, another for no. It is a method for tapping into a level of self that 'knows'. The best crystal to choose for a pendulum is similar in size to the one you would wear. I believe it should also be very clear.

'Working' crystals

For use in the garden the more opaque or earthy types of crystal can provide good balance (see the colour illustration after page 106), the kind of stone you feel 'comfortable' with. These stones are more connected to the earth by the very nature of their opaqueness. White quartz and smoky quartz are very suitable. Rose quartz provides a very soft, gentle, nurturing energy for earth healing. If your garden needs more stimulating energy then choose a clear crystal. Aquamarine and amethyst are both wonderful stones for water gardens, where pools, streams and ponds are present.

You may decide that you want more than one crystal. In fact if you intend working with them, as I myself do as part of my healing and therapy sessions, it is a good idea to separate your working crystals from those that are personal – personal crystals being ones that you do not let anyone else touch.

My 'working' crystals are freely available for anyone to feel, handle, even borrow. I have other crystals that are as connected to me as a part of my own body, and whose vibrational field I do not want affected by any outside influence.

The reason for this is that the 'atoms in motion', the molecular structure of all matter around us, can be affected by thought, emotion, and happenings. Crystals are similarly affected. This explains why a psychic can walk into a room and 'read', or sense, what has taken place previously in it. Memories are impregnated into the walls, floor, ceiling. This 'absorption' causes a shift, a movement. Personal possessions, also, absorb the qualities and vibrations of their owners.

In the same way crystals are affected by the energy of those who have handled them. They need to be cleansed and cleared in order to make them truly yours.

Cleansing crystals

There are various methods for doing this. For me the best is to soak a crystal for 36–70 hours (36 hours minimum and 70 maximum) in a solution of pure seawater, or half a teaspoon of sea-salt per pint of warm water, in a bowl in which the crystal is fully submerged.

Water method

To clean the 5–6in crystals which I use during my healing sessions, I would put two to three teapoons of sea-salt in four to six pints of warm water so as fully to submerge these crystals. I prefer to use seawater if I can get it. (The crystals prefer it too!)

Sea (or rock) salt is available in health food shops and most supermarkets. As the vessel for soaking the crystals, china,

Cleansing a newly-acquired crystal with sea salt

pottery, glass – even the kitchen sink – are better than plastic. If you use sea-salt, be sure to add enough warm water to dissolve it completely, then add cold water. Do not put your crystals into hot water or you may damage them.

Using the breath

Another way of cleansing is with the breath. You must take a moment to attune yourself to your own 'I am' centre, or higher consciousness, by thinking of your crown chakra. This is the energy point at the top of your head around your crown. Imagine a flame like a candle flame there. Then, using your imagination, bring the flame down into your mouth and breathe out, exhaling sharply and emphatically into the crystal.

If it is a small crystal, one breath is enough. If it is bigger, you

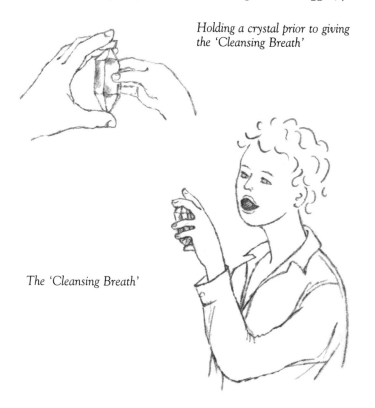

Holding a crystal prior to giving the 'Cleansing Breath'

The 'Cleansing Breath'

will need to breathe into each 'facet' or side of the crystal. It is a good idea mentally to affirm 'I cleanse this crystal in the light and love of God' (or the Holy Spirit or Buddha – or whatever divinity matters to you personally.) 'Let only that which is of its own consciousness remain, and let all else be transformed into the light.'

If you have 'programmed' or 'empowered' the crystal – that is, directing it to a specific use – you might say, in the same way, 'In the name of love I clear and cleanse this crystal of everything save the programme and that which is of itself. I command any and all negativity to be taken to a place of light to be transformed.' It is important that you do this in your own way and with your own words.

How often should this cleansing be done? Always when you first acquire a crystal and every few weeks thereafter. If you are using your crystals to work with many different people you will need to cleanse them more often. If only you are handling your crystal you will need to clean it less frequently.

You can also wash crystals in running water – again while mentally commanding the release of anything inappropriate and its subsequent transformation. If, at the moment of wanting to cleanse your crystals, you have no access to running water, you can still visualize cleansing them. Imagine that sun, light, water, and so on pours on to them in exactly the same way that you would cleanse them.

When my crystals want a bath I usually learn about it in my dreams. I use both the soaking and breathing techniques. The soaking compares to the breathing as our bath compares to a shower. Sometimes the crystals, like us, prefer a long relaxing bath.

Once removed from the earth crystals enjoy sunlight and air. Most of mine are scattered on window sills, tables, bookcases. A few prefer being wrapped up in black silk or velvet – any natural fabric is compatible with crystal.

The more time you spend with your crystals, the more you love and appreciate them, the greater the response. As a result you will sense what they like and dislike, what makes them happy. Many of my crystals have developed rainbows, in the same way that flowers blossom and bloom if acknowledged and appreciated. In Australia I borrowed a friend's crystal to show on a television show. The interviewer held it towards the camera and for almost one minute the crystal filled the screen. It was a sensation. People began phoning the station to talk about its effect on them. When I returned the crystal to its owner he could not believe how the crystal had changed. It gleamed and shone as if filled with light from inside. The crystal had become a television star and loved it.

Crystals are here to stimulate us to become stars — stars of our own lives, our own movies. Enjoy them, love them and do not be afraid to experiment with them.

Aurichalchite (see photo opposite page 107) is used in healing in the same way as other blue stones. It is a stone of harmony and balance which calms and helps to clarify thought.

The practical benefits of crystal power

Crystals have become big business. Publicity has stimulated such interest that people are using them almost like a magic potion to cure everything from alcoholism to athlete's foot to better communication with the mice who raid the kitchen at night. Interior decorators design homes around them. In America businessmen install them in corporate offices. The Earth on which we live is like a giant crystal and the physical body is a complex matrix of liquid crystals.

Last year a Crystal Congress was held in Los Angeles with an audiovisual hook-up of two of the world's biggest crystals, one in Moscow, one in LA. Crystal people gathered in America and Russia in an exchange to promote peace, friendship and co-operation.

Crystals played a part in the technology that sent rockets to the moon. They are being planted in ancient sacred sites, such as stone circles etc, throughout the world to heal the Earth. Without crystals we would not have the communication system we have today. They even counteract the low-frequency vibration from microwave ovens and television sets. Medical and scientific reports confirm their healing properties; healers, therapists, farmers, fortune-tellers, witches and wise men have all used crystals for eons of time. They can be used safely and successfully with adults, children, plants, animals and food. Even machinery appears to respond to crystal power!

Using crystals in your daily life

A friend of mine put a crystal under the bonnet of her car when it developed engine trouble on an isolated stretch of road. The car livened up enough to get her to a garage about eighty miles away. She never drives anywhere now without placing a crystal under the car bonnet. Most of her friends do the same thing, having found that it works.

Crystals have myriad uses. They take in one energy and give it out as another. If you put electricity into a crystal it gives off vibration. If you hit or squeeze a crystal it gives off light and electricity. In each case the energy that goes in is not the same energy that comes out. It has been transformed. Each crystal has a unique force field that can stimulate growth, penetrate and affect every level of consciousness and matter. It is responsive, transforming and regenerating. Crystals stabilize the emotional body, while energizing the physical. Crystal power improves the ability to dream and meditate, helps develop intuition and the psychic sense, increases the power of the brain to learn, and to remember what is learned.

So vast is the potential of crystals that it can be extremely difficult to know how to start using them.

Getting to know your crystal

The very first step, after acquiring and cleansing your crystal, is to get to know it. You do this in exactly the same way you get to know a new person in your life. You spend time with it. Feel it, hold it, stroke it. Don't be afraid of it. I opened a box of crystals for a workshop in South Africa, and stroked three crystals that immediately lost their cloudiness and lit up. I was so excited I spent almost the entire night stroking about 250 crystals, but they did not respond in the same way. Even when understanding a little of the science behind crystals, there is still the unknowable X-factor. It was a wondrous and magical experience.

The more time you spend with your crystal the greater your communication and rapport with it will be. Listen to it — crystals do actually sing. Ask the crystal how it can best help

Amethyst (note the rainbow at tip)

Agate Geode

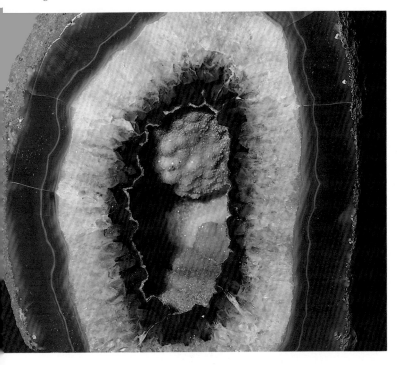

Black Quartz Crystal

Malachite

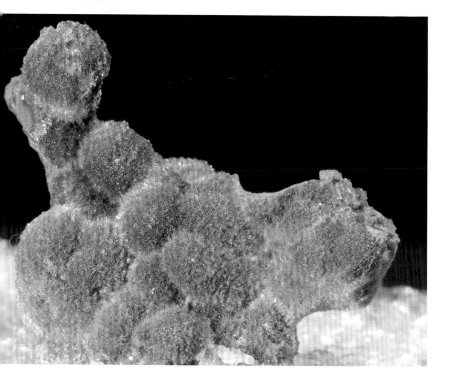

Quartz Crystals and Chalcopyrite

Sceptre Quartz Crystals

Rhodochrosite

Aurichalchite

you and make a note of anything that comes into your mind. It is a telepathic communication and you may not get the answer immediately. If it is a small crystal carry it, wear it or put it in your pocket. I know a number of girls who slip small crystals into their bras. You may need to wrap it in a small piece of silk or velvet, or even a piece of cotton duster. Remember to use natural fibre rather than synthetic. This is to protect the crystal when you carry it. Otherwise most crystals enjoy sun and light.

If your crystal is too big to wear or carry in your clothing keep it within a 3 ft radius. Within this radius its force-field will continue to affect you whether you place it on a kitchen table, an office desk or inside a handbag.

Crystals placed on a desk for inspiration

Another method of attuning to a crystal is to hold it in the right hand and touch the point to the palm of the left hand. Then move the crystal 1½–2in out from the palm whilst rotating it clockwise, until you feel a pull or tightening. The feeling is a little like when you tighten a screw-top jar and it suddenly 'takes'. This shows you are aligning with the crystal. It helps if you shake your hands vigorously first, place the palms together and slowly move them apart until you feel the magnetic pull from palm to palm stop. If you do this two or three times, before any hand-sensing or scanning technique, you stimulate the energy current in the hands, thereby increasing their sensitivity. Holding your hands two to three inches above a crystal cluster will also recharge the electro-magnetic energy in the hands. Placing a crystal on a cluster for approximately twelve hours is also a method for cleaning and recharging the crystal.

The Cherokee Indians have a beautiful ritual for bonding with their crystals. First they run them under fresh water for twenty minutes. Then they give thanks to the crystals, for coming into their lives, by placing a few grains of tobacco or corn on them. Next they burn some cedar and pass the crystals through the smoke, letting the fragrance go through every part. After this they wrap the crystals in a soft dark cloth and keep them covered, from new moon to full moon. During this time, every night, while keeping them wrapped, they hold them near their hearts and communicate with them, sing to them and love them.

Whether you use a ritual such as this, or simply take the crystal to bed with you every night, the main purpose is to become friends. Once you and your crystal are comfortable with each other you can start using it. Many people decide that the first step in using a crystal is to wear it around their neck (see page 96.) If you decide to wear your crystal do not wear it all the time.

Most crystals, except double-terminated ones, are one-pointed and are usually worn with the point downwards. Marcel Vogel, the crystallographer and scientist, found that the energy which crystals constantly emit causes a leak, over a period of time, in the wearer's energy field. If you need grounding then to wear a crystal point downwards can help stabilize you. For the purpose of meditation, visualization, or awakening intuitive faculties, it is better to wear the crystal point upwards. However in both cases you are working with a force which, like medicine, you can use to excess. Sometimes wear it, sometimes do not.

Muscle-testing with crystals

You may like to test the effect of crystal power with kinesiology, or Touch for Health, as it is often called. In my workshops I frequently use one of the kinesiology techniques to test the effect of a crystal on someone's body. This is also a very simple method for checking for food allergies, how negative and positive thinking affects our bodies, whether certain colours are good or bad for us, or if the jewellery worn needs to be cleared because it is weakening the muscles. Kinesiology is basically a method for testing and strengthening muscles. It allows the subconscious to tell us what is good or bad for us. The process requires two people – one the 'patient' or 'subject', the other the tester.

The subject stands erect and raises his right arm to a horizontal position. With his right hand the tester pushes gently down, with his palm open, on the subject's extended arm, to

Muscle testing with a crystal

feel the normal muscle strength. Then the subject holds whatever needs to be tested in his left hand long enough for the subconscious to assess whether it is beneficial or not. The tester then reassesses the subject's muscle power by again pushing down on the raised right arm. If what is being tested is good the subject's muscle power will increase in strength. Sometimes the arm will almost bounce up slightly. If what is being tested is bad for the subject his muscle power will diminish, causing his arm to drop or give. In other words a potentially harmful substance will weaken the muscle and a beneficial one will strengthen it, or will have no effect.

When doing this I have often discovered that people wear, eat or use things that have a negative or weakening effect. Even more startling is to see the way our thoughts affect ourselves and others. In experiments I asked one group to send nasty thoughts (mentally saying I do not like you etc.) towards a volunteer. The muscle test afterwards showed such weakness that the volunteer could barely lift his arm. This was done without telling him, until afterwards, what was being directed towards him. As soon as we sent affirmations of love and liking towards the volunteer his arm bounced back and his whole body stance changed. I have done this all over the world literally hundreds of times and the results prove the power of thoughts to strengthen or weaken anyone around us. In testing we also found that thinking, 'I don't like you' about someone else, weakens not only their muscles but ours too.

Crystals can help strengthen our muscles so that the bombardment of other people's thoughts and emotions do not affect us in the same way. However, how much better if we could start looking at each other with love and acceptance instead of hostile judgement and criticism. 'I love you even if I do not always like what you do' is a very muscle-strengthening attitude.

Breathing good, bad, happy and sad thoughts into crystals was also tested with kinesiology in a crystal workshop. One person thought of the worst time of her life and pulsed the breath into a crystal. Another thought of the most wonderful experience in his life and breathed that into a different crystal. Each person, when holding the crystal containing the 'worst time' thought, experienced extreme muscle weakness. The crystal impregnated with joy, stimulated strong muscle response in everyone. People were tested individually and were told only that different crystals were being used for an experiment. This was so that the conscious mind did not interfere. When doing any experiment like this make sure you clear the crystal afterwards.

Kinesiology can be used when choosing a crystal. It can also be used to test the efficiency of various methods of working with crystals. Kinesiology is far more than this one technique. Many books have been written about it and there are wonderful practitioners available if you decide you would like to know more about it.

In every crystal workshop I have given during the past ten years I have found at least one person whose muscles appeared weak when tested because of a crystal worn around his neck. Often the crystal had been set in such a way that it could not breathe and needed to be reset. Sometimes cleansing alone was enough. Marcel Vogel suggests that to cleanse a crystal you should hold the two ends, point up, between the thumb and forefinger of the left hand, while holding the right-hand thumb and finger on two faces (see illustration). Mentally think of cleaning the crystal of anything that is not for your highest good and pulse your breath into the crystal on every side. To complete this process draw another breath while thinking love, joy, happiness and breathe into the crystal again.

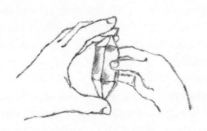

Hold a crystal this way prior to programming it

This can be done to programme a crystal if you have a project to accomplish or a physical problem you need help with. Breathe the thought of what you want accomplished into the crystal and its energy will help you. Because a crystal responds to our intent and need, we automatically programme it but sometimes it can give one a feeling of extra support to do this.

This cleansing can be done with any other gems or jewellery too. Muscle test the possible harmful or beneficial effect before and after clearing. Occasionally a gem needs this to be done more than once. The muscle test will tell you when the cleansing is complete.

I once had a small diamond-shaped crystal in an unusual formation with a second diamond-shaped crystal growing out of it. I had it made into a ring. I explained at length to the jeweller the angle at which I wanted it set, stressing that the setting must hold the stone in position without in any way closing it off. I was leaving the country so I paid a large sum and arranged for the ring to be posted to me when ready. I received the parcel with great joy and excitement. I opened it, and found to my shock and disappointment that he had done everything wrong. I put the ring on my finger and experienced excruciating pain. It felt as if the crystal was suffering in the same way as if I had jammed my finger in a door. I took it at once to another jeweller to be freed. It needed a lot of healing, cleansing and loving but it is now happy and free of any setting.

Simple self-healing with crystals

Aside from wearing a crystal there are dozens of simple ways to improve the quality of life with crystal power. They are wonderful tools for self-healing. Initially the easiest method for self-healing is to carry the crystal around with you and take it to bed with you.

When the battery runs down in a radio it produces an imperfect sound. When the battery in your body runs down or we become under par, we don't think properly, we become tired and unco-ordinated. We too give off a discordant sound. The crystal recharges the battery. It regenerates the cells of the body and activates the circulation. It stimulates the lymphatic and glandular system of the body and I have found it particularly helpful with post-operative surgery. Imbalance in the cells causes static in the energy field or aura. Imbalance in the aura causes disease in the cells. Crystal energy can penetrate and aid the dispersion of this static or imbalance.

Two friends who recently underwent major surgery not only recovered far quicker than is normal but had no problem with an angry or slow-healing scar. They both kept crystals beside and inside the bed. Inge asked her crystals as much as she asked the doctors, whether the hospital medicine was appropriate or not. She also crystallized everything she was told to swallow by whirling her crystal over the top first. She went into hospital on the verge of death and came out looking twenty years younger. Crystals did not do it for her but they certainly helped. Jill Ireland used crystal therapy as part of her treatment for cancer.

Crystals are good for back and bone ache – lower back pain especially. If I have a problem, whether it be a headache, stomach-ache, twisted ankle or knee, I put the crystal in my bed, even under my pillow. By morning I'll find it alongside the part of my body that needs healing. A girl in a workshop, who borrowed a crystal for the night, got the shock of her life when she found the crystal wedged against one ear in the morning. She had gone to sleep resting her feet on it. It turned out that she had been having problems with her ears which the crystal obviously tuned into. After that she bought her own crystal and used it to heal herself.

You will create a wonderfully healing atmosphere in which to sleep if you place a crystal on the floor pointing inwards towards the bed. For this I find 3–12in crystals the most effective – longer if you have them. I used this technique with a friend who had badly scalded her arm and was in great pain. Within two days the pain and the burn was infinitely better.

It is important to vary what you do with your crystals. Sometimes I put mine under my pillow. Sometimes I'll leave it in the bed. At other times I will simply place it on the bedside table. I keep a crystal on my desk and even in my bathroom. If I'm having friends to dinner I decorate the table with candles and crystals. They spark up the atmosphere and stimulate conversation. I always travel and work with them. I place crystals around the people I'm doing healing sessions with. They can be a therapy on their own or an adjunct to any other therapy. I sit with a crystal on my lap when I'm reading or knitting.

I was doing this on a plane when the woman next to me clutched my arm and told me how frightened she was that the plane would crash. She hated travelling, whereas her husband loved it. Since retiring he was continually arriving home with tickets to unexpected, faraway places, not realizing that his wife died a thousand deaths each time. I suggested that she held my crystal and take some deep slow breaths. I held her hand for a time and then discovered that she had fallen asleep. She slept peacefully for most of the journey and awoke pink-faced, glowing, amazed at what had occurred. She wanted to go straight from the airport to the nearest rock or crystal shop to buy one for herself.

Another time I was flying, with some very large quartz crystals and various other rocks and stones. My luggage weighed 98 kilos. All the way to the airport I thought light thoughts towards the unknown person who would weigh my bags. As they were deposited on the scale, with my hands in my pockets and a crystal in each hand, I continued to think light.

'How many of you are travelling?' asked the airline official.

'Just one,' I replied.

He looked from me to the scale and repeated, 'How many −?'

'Just one,' I said.

Before he could say another word the stapler he held in his hand exploded, shooting staples all over the counter. Looking extremely confused he picked them up while I waited. He turned back to the scale and opened his mouth to speak when the stapler exploded a second time.

As he gathered the staples together again I said, 'Is this a busy flight or a light flight?'

Slightly dazed he replied, 'It's a light flight. A light, light, *light* flight. The lightest flight I've ever experienced since working in this airport. Gee, is this a light flight . . .'

He passed all my luggage through. He never said, 'Madam, your luggage is grossly overweight, you must pay extra.' He got the message that it was a light flight and so was my luggage. Baggage attendants often say to me, 'You must have rocks in here,' and when I agree they smile disbelievingly.

Re-energizing and rejuvenating your body with crystal power

Travelling as much as I do I frequently feel in need of a quick recharge. A simple way to do this involves holding a crystal in each hand. The crystal in your left hand must point towards the wrist, the one in your right hand must point away from the wrist. Five or ten minutes of this exercise will spark you up as if you had plugged yourself into an electric socket. If you have only one crystal hold it first in one hand and then in the other. It does not matter which hand you use first, but remember that you should still point the crystal in your left hand towards your wrist and the crystal in your right hand away from your wrist.

To re-charge yourself, put one crystal in the right hand with the point facing away from the palm, towards the fingertips, and put another crystal in the left hand with the point facing up towards the elbow.

Another method of re-charging is to hold a crystal over the thymus

(remember that the left hand receives energy and the right hand transmits it, (see p.89).

Another powerful re-energizing technique is to place a crystal on the thymus. This is a gland approximately 4½in down from the throat, near the sternum. The ancient Greeks thought of the thymus as the fire centre from which all else flowed. Some religions refer to it as the 'witness' area, believing it to be the seat of wisdom in the body.

If you place a crystal on this area for four or five minutes you will revitalize the thymus and in so doing will recharge the muscles and tissues throughout the body. The effectiveness of both these exercises can be checked with the muscle-testing technique from kinesiology.

When you begin to use crystals you will find that just as your body becomes healthier so too does your skin improve. Because crystals regenerate the cells of the body they appear to help smooth wrinkles and soften the skin. Most of the women I know, who have used a crystal for some time, have an inner glow and radiate well-being. Rose quartz is a beautiful crystal for the skin. Masks carved from rose quartz were applied as a beauty treatment in Egyptian times. A number of people prefer rose to clear quartz as it is more gentle and comforting. It is effective in treating menstrual pain and provides a nurturing energy at times when a woman feels vulnerable — after child-birth, during menopause or when a love affair ends. (Men, in this situation, find malachite helpful.) I have a big chunk of the most delicate, silvery pink, unpolished, rose quartz which I hold when I feel sad or when life seems to be falling apart. It is a crystal that gives support, helps one to feel better about oneself in a quiet, calm way. It absorbs emotional anguish.

To improve the skin you can use crystals internally or extern-ally. No, I do not mean swallow them! A beauty treatment that stimulates good skin tone is to hold the crystal lengthwise, in the right hand, about two inches away from the body, with the point of the crystal going towards the fingertips, and gently massage the energy field around the face. You do not touch the skin. You rotate the crystal's length clockwise in small circles. The area covered needs to extend about six inches from around the face and neck. This treatment should be done for approxi-mately twenty minutes. You can work on your own face but it is easier, and more enjoyable, if someone else does it for you. Maybe you can exchange with a friend.

When you've finished the massage brush the energy upwards from the neck to well above the head. You continue to hold the crystal lengthwise. This can be a very relaxing treatment for the whole body. I have used it many times in treating sleepless-ness or hypertension. Usually the patient falls asleep towards the end. If you do a full body crystal massage it is best if the patient lies on his stomach and start on the back first before turning over and working on the front. Again extend the area of massage at least six inches from the body line.

Whenever doing this type of crystal work it is important first to attune, or align, to the energy of the crystal by either holding it, or using one of the methods described earlier in the chapter. Do not get up immediately afterwards. The crystal energy needs to 'soak in'. You may also feel a little light-headed. If you have bad skin it is a good idea to massage with anti-clockwise movement *first*. This helps to draw impurities out of the skin. *Then* apply the clockwise movement. Skin problems usually reflect an inner sense of inadequacy. Rose quartz heals one's attitude to oneself, which in turn heals the skin. For this treatment clear quartz or rose quartz can be used.

I am sure the way my face healed after my accident was significantly helped by my use of crystals.

(a) Purifying a glass of water by placing a crystal in it

Purifying water with crystals

Another way of developing good skin tone is to put a crystal in the bath-water. Don't use boiling water in the bath as this would harm the crystal, but a temperature that is tolerable to you will be sufficient and will not damage it. Place the crystal in the bath ten minutes before you get in and soak in the water as long as you can. To help clear the body of toxicity drink a glass of crystallized water every morning when you wake up. You can crystallize water in two ways — three if you include directing the point of a crystal at, and touching, a glass of water. One method, which requires no effort, is to drop a crystal into a glass or jug of water overnight. This charges the water making it both cleansing and energizing.

The other is a little more involved but can be used to remove the taste of fluoridization, caffeine, tannin and preservative in fruit juices.

First, fill a container and sample the water (or whatever

liquid you decide to work on) so that you can test-taste the difference. Clasp the crystal around its middle, point down, in the left hand and hold it 1–2in above the container. Both ends of the crystal should be visible, so the crystal needs to be a minimum of 4–5in in length. With the right hand cupped, clasp the top end of the crystal and turn it anti-clockwise, while maintaining your firm grasp with the left hand, as if unscrewing a screw from its hole. Tighten your grip with the right hand and take away the left hand. With the crystal pressed firmly against the palm, rotate in a circular clockwise motion until you sense that it is enough. The anti-clockwise motion unlocks the structure of the liquid, the clockwise revitalizes it. Now taste and if you do not notice a difference continue until you do. Sometimes the water develops tiny, effervescent bubbles.

I have used the same technique with wine. It was highly

(b) Purifying water by the 'unscrewing' method: clasp the hand around the crystal so that both ends are showing, then 'unscrew' anti-clockwise with the right hand

successful in changing rough, cheap plonk into a smooth, mellow-tasting wine! If this sounds far-fetched I have just received a letter concerning four wines bottled by Sycamore Creek Vineyards in California. They used a process developed by Marcel Vogel and when testing the result more than 90 per cent of the people polled in the tasting room could detect differences between the 'ordinary' wine and the Vogel-processed wine. The tasters were told nothing about the Vogel processing and were simply asked if they could detect differences in the wine.

Terry Parks, the owner of Sycamore Creek Vineyards, says,

> In my own estimation, as a wine-maker, the process smoothes and structures the wine and increases its palatability. It seems to eliminate or reduce the need for extensive bottle ageing. As such, this process may well have significant commercial applications in the wine industry and may carry over into other beverage industries.

Energizing food and plants with crystals

A crystal cluster, or a single crystal, placed in the kitchen near food being prepared will recharge the life-force of the food. To keep fruit and vegetables fresh, place crystals around them.

Crystals placed around fruit/vegetables keep them fresh for longer

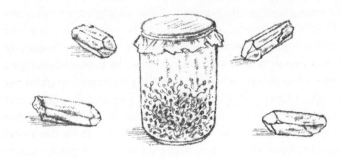

Crystals can be used to make seeds sprout faster

Crystal power works like pyramid energy and can be used for milk and all kinds of food. If growing seeds, such as alfalfa or mung beans place a crystal near the jar. They will develop faster and stronger. (For that matter so will seeds for the garden.) Dropping a crystal into a vase of flowers, or putting a crystal near the vase or alongside a plant, will prolong their life and health. Some stores now sell cut flowers with a crystal. Sometimes an indoor plant will grow towards the crystal instead of towards the sun. For stimulating growth the crystal should be directed towards the target.

If a person preparing food is in a bad temper the energy field around the food will take on a greyish tinge. In restaurants I

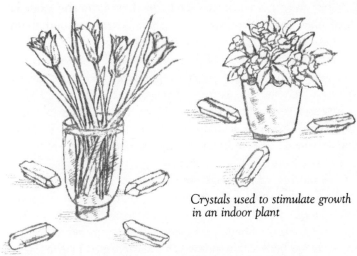

Crystals used to stimulate growth in an indoor plant

Crystals in or around a vase will prolong the life of cut flowers

sometimes use a small crystal to circle the plate and pull away the grey.

Microwave ovens emit what is called LFV, or low frequency vibration, so do television sets and most electronic equipment. Research has shown that LFV can have a depressing effect on one's emotions, just as fluorescent and artificial lighting has been proved to be physically detrimental. In fact tests, involving children who had learning difficulties or behavioural problems, showed that when placed in rooms with natural lighting the children's behaviour and ability to learn improved. Keeping a crystal near your television set or microwave oven counteracts the negative effects of LFV. Giving crystals to children also helps counteract the stresses they are subjected to.

Crystals and dreaming

Children love crystals. Crystals love children. During a series of children's workshops, in which we explored dreams, meditation, colour, sound, movement, simple numerology and astrology, and many other wonderful and creative fields, we also explored the wonders of crystals. Each child, having chosen his own crystal, was encouraged to experiment with it and soon became adept at scanning the aura, or energy field, around plants and people and giving simple crystal treatments. Crystals were used to grow seeds, heal pets, soothe parental headaches, help dreaming and solve problems. After the children had used crystals for a time parents began to comment on how much more loving and responsive they were. The children discussed each other's nightmares and were encouraged to face the monster, or threat, in a nightmare, while pointing a crystal at it until it shrank to a manageable size. They could then pick it up and make friends with it. (This comes from a Senoi Indian technique when children are taught to point a finger at the threat.)

One mother was so impressed by her son's achievement, in dealing with his monster in this way, that she used the same formula when dealing with a dream monster of her own. Her monster took the shape of a grotesque creature who climbed through her bedroom window. The fear aroused by the dream woke her up but after a few moments she remembered what her son had told her. Closing her eyes she mentally recreated the creature climbing through her window and imagined pointing a crystal at him. To her amazement he shrank into a knee-high dwarf, who ran out of the room. A week later a real man climbed through her window, with a knife, and threatened to kill her. Cowering in her bed, convinced she was about to die, she suddenly remembered the dream. She summoned all her courage, sat up, pointed a finger at him, and said, 'Drop that knife and leave this room immediately.'

The man was so shocked that he dropped the knife and ran away. She mentally pointed the crystal at him in exactly the same way as she had when dealing with the nightmare. Obviously the dream had been pre-cognitive and gave her the power to act when it was necessary. Dreams have an energy. If we work with them we pull the energy of the dream into everyday life in a way that can be totally transforming and, as in her case, life-saving.

Dreams heal and balance us — without them we would become psychotic — although many people do not remember their dreams. Crystals can both stimulate the ability to dream and help us to remember what we dream. Put a crystal under your pillow at night and see what happens. Breathe into it while saying, 'I want to dream and I want to remember my dream.' When you wake up in the morning the first thing you must do is write the dream down. Take the crystal in your left hand and ask it to help you to remember. Even if you do not initially remember the whole dream it is important to write something down — the feeling or colour in the dream, the feeling you wake up with. If you have not paid attention to dreams before you need to encourage the part of your mind that talks to you through dreams. Dreams put you in touch with your real self, your wise self and can be a source of great inspiration and information. The Egyptians used dreams to ask questions about every facet of life. We can do the same thing today.

Crystals, children and the natural world

Crystals placed in any room will create an atmosphere of

harmony. A tiny crystal placed on a new-born baby's cradle – even an older baby's cot, as long as it is hidden and cannot be reached by the child (such as under the cot) – will engender a beautiful energy for the baby to sleep in. Crystals are beneficial in a room where a person is unwell or dying. They dissipate the aura of sickness and shadow. Once during a healing session with a young girl who had cancer, a tiny girl crawled into the room. This little tot who could not even stand up, stared at the crystals and immediately picked up the biggest. She brought it up to her forehead and then hitched up her clothes and rubbed it all over her tummy. She even used a clockwise movement! She then took it to her mother and gestured to her to do the same thing to her back. Like many of the children of today who are wise old souls she knew exactly what to do.

A doctor I know put a large crystal cluster in his waiting room. After about ten days he realized his patients were decreasing in number. Soon almost no one came to his surgery. He could not understand what was going on and worried how to pay his children's school fees. Gradually a different group of patients started coming to see him. He suddenly realized that the arrival of the quartz had cleared his practice of all the people who did not really need his treatment and opened a space for everyone who did. He discovered that with crystals his treatment appeared to work better and the patients recovered more quickly. He acquired more and more crystals and uses them as part of his practice now, as do many conventional or allopathic doctors in America.

In Sweden crystals are crushed, finely ground and put on the roads in icy conditions to prevent car accidents. Bio-dynamic farmers who farm organically, also taking account of seasonal and planetary changes, use crystals in the same way to regenerate the earth and stimulate plant growth. For eons of time ancient peoples believed that quartz, when crushed to sand and sprinkled on the soil, brought the energy of the stars down to earth and that whole crystals brought the energy of the stars into the soul.

An elderly friend of mine, Molly, has one of the most exquisite gardens I have ever seen. It is a small cottage-type garden with a profusion of flowers, full of fragrance and dazzling colour. It is a place that fills one with peace and delight. Molly is now in her eighties and so I asked her if she had some help. 'Yes,' she said, 'Let me introduce you to my help.' She pushed me on to my knees to look closely at the beds. I discovered all sorts of crystals tucked under the plants and evidently working away as hard as they could, the beauty of the garden demonstrating their efficacy.

An American Indian technique for energizing a garden, or a plot of land, is to square it off and plant a crystal in each corner with a fifth crystal in the centre. This central crystal should be planted pointing down into the earth. The four corner cyrstals must be directed towards the crystal in the centre. This creates a current of energy similar to pyramid energy. If for some reason one area of the garden needs extra help, you remove and place in it one of the corner crystals, leaving it there for as long as is necessary. When the area has recovered, replace the crystal in the corner. From personal experience I know this works.

In many parts of the world people are experimenting with plants and crystals. In New Mexico, USA one experiment, with cactus plants, showed that a cactus grown with a crystal pointed at it grew 20 cm (8 inches) longer than one without.

Crystals affect living things because they have similar internal structures to living organisms. Crystals grow and develop to maturity like living things. A crystal is a form of life we do not fully understand but one that is as capable of transforming life in the twenty-first century as electricity or the Industrial Revolution did in the past.

Most animals are extremely responsive to crystals. To give cats and dogs, or any animal, water in which a crystal has soaked, instead of untreated water, recharges their energy as much as it does ours. With the co-operation of friends I tested crystal-charged and untreated water over a period of time and found that the animals automatically drank the crystal-charged water and left the other untouched. I was looking after a friend's dog when I decided to soak in a crystallized bath. The dog came into the bathroom, looked over the side of the bath and, to my amazement, jumped in too.

I discovered that some animals are very sensitive to a crystal held over different parts of the body and some appear not to notice. A guide dog who came to a crystal seminar with its owner was extremely sensitive and with the crystal above him the hairs on his head almost stood on end. I know many people who use a crystal as an adjunct to veterinary treatment and find that it speeds up the cure. A vet in South Africa treats all her animal prescriptions with crystals before she administers them. They were certainly a great help in nurturing my relationship with my parrot.

Much earlier I had another parrot who flew into a tree in the garden and would not come down. Day after day I wept and waited, begged and implored him to come home but to no avail. Eventually I used a crystal, almost like a telephone, to ask the Guardian Angel of Parrots either to bring him back or show me that he would be all right. (A bird that has been a pet is often attacked by wild birds and doesn't know how to defend himself.) Five hours later my parrot landed on the lowest branch of a tree just outside the window. He was within a hand's grasp and he 'spoke' to me. In my head it felt as if he were saying, 'I'm OK. Don't worry.'

After ten minutes he very slowly lifted into the air above my head and flew in circles around me, hovering. A sudden squawk drew my eye back to the tree and on the same branch stood an almost identical bird. The second, more mature, bird also spoke telepathically and seemed to say, 'Your prayers have been answered. I'm here and I will look after him. You shouldn't worry.' He spent approximately eight minutes ensuring that I understood, then he too lifted off the branch and flew in slow circles around my head.

After this, both birds lined up alongside each other and flew away. I felt as if a glass shard of pain was squeezed out of my heart and I blessed them and released him. This happened in Australia and for as long as I stayed in the same house they came back once a week to say hello.

Crystal messages

You can use the crystal telephone to tune in, or send loving thoughts, to family and friends who are away from home. Hold the crystal while thinking of them and what you would like to say, then pulse the message into the crystal with the breath. This is a wonderful method to keep in touch with children, if you are apart through divorce or separation. It keeps the bond between you alive, like watering a flower. To place a photograph under a crystal, in order to send love and healing, is also simple and effective. The crystal amplifies and focusses what you can do with your mind and your energy. Crystals can be used like this to dissolve and solve communication problems. Penetration twins (see p.94) are ideal for getting to the root of a problem.

Whether male or female, we all have an inner counterpart of the opposite sex. The inner woman in a man allows him to relate to, and express, his feelings and intuition. The inner man in a woman stimulates her to be active and creative. Frequently, as a result of bad or non-existent male/female role-models in childhood, our relationship with our inner man and woman can be non-supportive or non-functioning. A crystal worn or carried on the left side of the body will help open us to the inner woman — the receptive, listening, emotional part of us. A crystal kept on the right side of the body will activate greater strength, confidence, assertiveness — the ability to act out the ideas of ones's life. Aside from this, if a particular situation requires you to listen, or 'take something in', wear the crystal on the left. If you need to give a talk or speak out wear it on the right. If you need to study, and store the memory of what you are studying, hold the crystal in your left hand while you do it. Try this both with the crystal and without and discover for yourself the difference it makes.

Meditating with crystals

For meditation it is also advisable to hold the crystal in the left hand. This awakens the part of your mind that enjoys the expansion of consciousness that meditation brings. You will find it much easier if you take some deep breaths, and tense and relax your muscles first. This helps the body to release tension and co-operate with, rather than fight against, you. Unless you can sit cross-legged in a yoga position it is best to sit in a chair with your feet on the ground, or lie down. When I started learning to meditate I thought I had to sit cross-legged. I ached, twitched, got cramp and nearly went crazy with discomfort. I now believe that it is vital to be comfortable.

You may like to light a candle to create a peaceful atmosphere around you. It is better at the beginning to practise
being still and relaxed for 5–10 minutes, or as long as you feel
comfortable, rather than trying complicated techniques.
Imagining or remembering a place in which you've been happy
is a good place to start and automatically relaxes the body.
There are many forms of meditation but they all require that
the mind be still. Once you are able to do this, experiment to
find the type of meditation that suits you best. Each time you
use the crystal it will store the memory of the meditation and
help to trigger your mind more quickly into this still space.

Another crystal meditation can be done by holding the
crystal in both hands. Bring the crystal to the level of the heart,
look at it, breathe with it, align with it. Then close your eyes
and gently touch the centre of your forehead with the crystal.
Keep it there for a moment or two then, with your eyes closed,
gently place the crystal in front of you or on your lap. Your
hands should be relaxed at your side or in your lap. Now see the
crystal again in your mind's eye, imagine it, remember it. Then
visualize your consciousness entering the crystal and imagine
being inside it for as long as you wish. When you feel ready to
withdraw from the crystal do it gently, gradually and slowly
open your eyes. You will feel light, renewed, revitalized. It is
wiser not to get up at once but to give yourself a few minutes
to absorb the experience.

Despite the many uses to which crystals can be put they are
not a substitute for what you can do with your own creativity
and power. Neither are crystals a substitute for orthodox
medical treatment, proper breathing and diet, exercise and
healthy living generally. They assist by focussing and
amplifying what we can do with our minds, energies and
willpower. Our bodies are controlled and regulated by different
energies pulsating through them. Crystals are tools to work
with and influence these energies. In addition they create a
harmonious environment where mind and feeling, body and
spirit function better.

Healing with crystals and stones

The art of healing has a history even more ancient and rich in magic than that of gems and jewels. The Etruscans projected their consciousness inside the patient and literally became one with him for the duration of the healing. The physicians of Petra in Ancient Greece (now Jordan) knew how to apply hot and cold sand and water and used mind control to cause the temperature change. These early physicians sometimes literally gave their lives to their patients by living with them in remote caves outside the city until recovery or death took place. Anyone sick or diseased was sent away and if the disease was infectious the physician often died too. Sometimes this willingness to give one's life for the patient was enough to result in instant cure and many miraculous healings took place. However, these physicians also knew how to work with universal laws, having undergone rigorous training.

In the Ancient Egyptian temples snakes were used to train the initiate how to understand, control and direct the flow of energy in the body. Part of the training involved mastering the mind in order to control the snakes themselves. Initiates were tested by being shut in a chamber with snakes – if they could not mesmerize the snakes they died. Snakes shed their skins and so were also symbols of transformation, the ability to shed the past and move into the new.

All ancient healers understood that you cannot treat the body alone, each healer-physician was considered to be a healer-teacher-priest. As humanity became increasingly sophisticated, the mind was separated from the spirit and both mind and spirit were separated from the body. The teacher taught, the physician treated the symptom (leaving the psyche to the psychiatrist), while the priest attempted to develop the spirit. It does not work. For complete healing on every level to take place, we need to know that illness exists first in the non-physical realm of mind and feeling, even spirit sometimes, before it manifests in the body. It is essential that we examine the whole person, treat the cause and not just the symptom.

Today this is what Holistic healing is all about. It is both whole and Holy. Crystals play an invaluable role in holistic healing because of their capacity to affect every level of vibration as well as every aspect of a personality, whether mental, physical, emotional or spiritual. Every part of us connects to every other part. If the lines get crossed, or the battery runs down, energy cannot flow. This creates disease, or a block or barrier, which crystal energy is able to pulse through and break down. Crystals will possibly become the surgical instruments of the future because of their capacity to penetrate both physique and psyche.

Today there are literally hundreds of both alternative and allopathic (or traditional medical) methods of healing. We have psychic surgery (the use of the mind to perform surgical operations) practised in the Philippines; heart and lung transplants are carried out all over the world; and homeopathy, acupuncture, aromatherapy massage and postural integration, and bach flower remedies, to name but a few, are all available to us. We also have the power to heal ourselves.

Healing power

The power to heal is the power to control and move energy, whether it be from one part of the body to another part or from one person to another person. Any therapy involves an exchange between two or more people. This exchange may involve touch, thought, a prescribed dose or any number of techniques. It works better when we believe in the particular therapy being used, and better still when used in conjunction with crystals.

However, as with crystals, there is an X-factor to healing. This X-factor is love or genuine caring.

When the American mystic and teacher Paul Solomon first developed his psychic ability, which is similar to that of Edgar Cayce, he asked his inner teaching source, 'What is it that makes a good healer? Can anyone be a healer?' He was told to go out and buy some tomato plants, to love some, hate some and ignore some. They all died. He did this many times and they all died. He became tired of killing tomato plants and decided not to be a healer. Six months later he accompanied a friend, who wanted to buy his wife a present of some cactus plants. Paul suddenly decided to try the experiment again, using cactus instead of tomato. He bought thirty plants and put them in his garden. This time he found himself responding to some plants more than others. He genuinely cared about them and they grew stronger and quicker than the others. He practised with orchids and found that the ones to which he genuinely responded emitted a fragrance. The others did not. He realized that his experiment to love, hate and ignore tomato plants had been purely an experiment. There was no genuine feeling, the plants sensed this and died. The X-factor was love and caring.

Dolores Kreiger, when training nurses in New York, became curious about what makes a good nurse and set up a programme of research to find out. She looked at religious background, age, efficiency, job satisfaction and interest, home life, hobbies, whether they were married or single, and many other factors. She could find no common denominator and put aside the research. One day a nurse, who had been away for a few days, came into her office and said, 'Mrs Kreiger, before I see her I want to know how Mrs X is getting on after her operation. I really care about her.' Dolores Kreiger suddenly realized that her research had never questioned the caring aspect. She reopened her files and continued the research. She discovered that in nearly every case when a patient and nurse cared about each other — when a patient looked forward to the arrival of the nurse on duty, when the nurse spent a little extra time with a patient, in a personal rather than a purely official way, the patient tended to recover. Even when nursing a terminal patient the quality of his life was markedly improved. Again the X-factor to being a good nurse is love and genuine concern.

I think it is vital to remember that anyone can be a healer, although some people have a greater power to transmit healing energy than others. The X-factor in sharing healing energy with another person is love and caring. We all have the capacity to love and care therefore we all have the capacity to heal, whether we use it or not.

Hairdressers heal by making their clients look and feel better. That is healing. Actors, singers, musicians and dancers are all healers. We are aroused by the passion of a performance to laughter and tears — enjoyment catalyses a chemical change in the body. We go in as one person and come out as another. Comedians are wonderful healers and laughter is a powerful medicine. Norman Cousins, former editor of the *New Yorker*, wrote a book called, *Anatomy of an Illness*, in which he describes how he used laughter to cure himself of what had been diagnosed as a fatal illness. He discovered that to watch a funny movie affected his body chemistry so much that he could delay taking a painkiller. As a result he prescribed for himself two hours of laughter, three times a day and eventually he totally recovered.

To be a good listener, to cook food that everyone enjoys, to create an atmosphere in a garden or house where people relax and feel comforted, is to be a healer. I stress this because so many people imagine it is something that only a select few can do. We can all do it if we want to, most of us are busy healing one another all the time without realizing that is what we are doing. Also pain stimulates the circulation and mind to move to the source of the pain. Energy follows thought and this is the body's way of (a) giving us a message of malfunction, and (b) ensuring that energy of life force moves to that spot. When we have no problem of our own we have an abundance of vitality to share with others.

This transference of vitality can be measured by wiring up both the healer and the patient to a brainwave instrument. Ten years ago I was involved with some experiments along these lines using Rose Gladden, the well-known British healer. Initially her brainwave rhythm had a different pattern to that of the patient. When she laid her hands on the patient's shoulders his brainwave pattern became identical to Rose's. We then used people in different areas of London and did the same thing except now Rose was merely given a name to mentally project healing towards that person. Exactly the same thing happened. Each person's brainwave mirrored that of Rose's, as the energy was transferred. It did not show exactly what happened but it showed that a transference took place.

Rose is a powerful healer. On another occasion the same experiments were carried out using 'ordinary' people — people who were not experienced healers. The process was slower and maybe the force was not so strong but the same blending of brainwave pattern between patient and healer began to take place. When crystals were used to assist in transmitting the energy, it was like switching on an electrical current. Both healers and patients experienced an enormous difference and everyone was re-energized. When I use crystals on someone else I usually feel recharged too.

Because crystals amplify thought it is important to clear yourself before you use them for healing, especially if you are feeling ill or depressed. A simple way to do this is to hold the crystal between thumb and forefinger, visualize the problem and discharge it by exhaling sharply into faces of the crystal — in other words you do it at least three times, turning the crystal in your hand. Affirm that any negativity be sent to the light where it can be transformed. Clear the crystal with the cleansing breath described in Chapter 5 and fill yourself and the crystal with light. This is a wonderful method to deal with aches and pains or anything that produces stress that can turn into distress. We need stress to activate us. When the accumulation of stresses becomes too great we move into distress which usually results in illness or accident. Use the crystal to discharge it — blow it away.

I do this sometimes even if I do not feel sick or worried. I simply ask to be cleansed by saying, 'Clear me of anything that is not appropriate for my highest good at this time. May it be removed from me and be taken to a place of light where it can be transformed.' It is important to clear the crystal and fill it and yourself with light afterwards. This can be used to remove the nitty-gritty stuff of the day, when you go to bed at night. It can be used in the morning when you wake up to free yourself of the night. It is a way of keeping yourself crystal clear.

As important as keeping yourself and your crystal clear is to remember not to force healing or think that you know best what the person needs at this time. It may not be what you think. He or she may need to increase his symptoms and get worse. S/he may need to understand the message of the illness, change an attitude, have a rest or a time to reflect. Perfect soul health is the ability of the body to experience the symptoms it most needs, to respond to those symptoms and then move on to some other experience. This may mean complete recovery or transition through death. In other words illness is always a teaching/learning experience, we often learn more through the limitations of illness that we do through health. Perfect soul health does not always mean a perfect physical body, often illness is a teaching for others as well as oneself.

Death and dying

Years ago Andrew Watson, a marvellous South African healer who now lives in Australia, told me how, when he first came to England, he was invited to a house to help a twenty-year-old girl who was dying of cancer. When he arrived he was shocked to see that the girl was in great distress, weighed only about three stone and was extremely ill. He was surprised that she was not in hospital. He discovered that her mother had organized a twenty-four-hour, round-the-clock, prayer-healing circle and to demonstrate her faith in this power believed she should remove her daughter from hospital.

Andrew prayed and meditated and had a vision of this girl walking down a road until she reached a barrier, similar to a railway level crossing. The barrier was formed by the prayers of the people in the healing circle. He explained to the mother that in trying to do the right thing they were in fact holding the girl back from what she was meant to do. After this the prayers were directed to what was best for the girl and within two days she died. Andrew then had a dream in which she came to him and thanked him for helping her. She appeared radiantly alive, healthy and free.

This story gave me the horrors when I heard it but it has always been a great help in illustrating what I mean when I say release any healing to the highest good of the person involved. Do not try to force recovery when it's not appropriate. Give your time, your thought, your love, yourself, as an instrument, and freely let it go.

One of the greatest fears we have as human beings is the fear of dying. This fear holds us back, limits us, prevents our fully exploring life. It is the fear of letting go, of change, of the unknown, of losing and leaving what is loved and familiar. It is the fear that I do not exist outside my physical body. For many people thoughts of death have all the terrors of a recurring nightmare. We lose out on life when we fear death. As Albert Schweitzer said, 'The real tragedy of life is what dies inside a man while he lives.'

Dying is very like a birth and birth is a natural process. Today death is rarely spoken of as natural and many regard it as injust punishment carried out by a vengeful God. If we thought of dying as the act of a person giving birth to another part of himself, instead of withering away into oblivion it would be far healthier.

We need to prepare for death in the same way as we prepare for a birth. Simple breathing, relaxation and meditational exercises are all wonderful preparation.

Meditation expands your awareness beyond the physical reality — a little like viewing the countryside from a helicopter instead of the family car. You gain a wider perspective. Dying allows the same expansion. Our feelings towards the dying person should be the same as the feelings we have for a dear friend, or family member, who goes to live in another land. We love them, we will miss them, but we wish them well as we wave them goodbye.

The room in which someone is dying should be full of light, flowers, candles, gentle music — an atmosphere of peace and upliftment rather than darkness and shadow, doom and despondency. It is the essence of the person you are now helping, to release from, rather than stay in, the body. A loving touch is also important at this time. In fact I would like to move on while having an aromatherapy massage, what a beautiful transition that would be. Crystals placed in the corners of the room, or on the floor around the bed, provide vitality for the etheric body during its birth into another dimension.

A man I know whose mother was paralysed for six months

before she died, spent day after day beseeching her for some sign of recognition. Then one day he realized that he was in some way keeping her anchored to this hospital room. He decided to change the atmosphere by bringing a number of crystals into the room. He then sat down next to the bed and, with a crystal in his hand, he looked at the ceiling and spoke to his mother. He thanked her for what she had been to him in his life, talked a little of their joys and sorrows, sang to her and then told her she was free to go whenever she wanted to. When he turned to the bed she quietly took a last breath and moved on.

The crystal is light, spirit and love; it represents another dimension and opens a way to move and communicate between earth and that dimension.

I was in South Africa and invited to a house where a young girl was dying of cancer. Her name was Cathy. Her left leg had been amputated and the cancer had spread into her neck and lungs. She was extremely uncomfortable despite injections for the pain.

I asked to be left alone with her and I placed crystals in the bed at each corner with the points directed towards her body. I gave her a crystal to hold in each hand and placed another on her thymus with the point directed towards her neck. After a prayer, asking for the presence of her own teachers and guides, I showed her a method of breathing, which relaxed her.

After breathing together for a time a great peace, almost like a tangible presence, entered the room. Cathy opened her eyes and I told her what the disease in her body was telling me about her life, her thoughts, feelings and problems. I asked whether she agreed with this and she confirmed everything I had said by talking about her life.

We spent five or six hours together and, through visualization, we healed situations with her father and mother, forgave whoever needed forgiving, explored death and dying as well as life and living. Crystals did not save her life. However, they helped to create an atmosphere that allowed feelings and thoughts to be looked at, spoken about, that had previously been hidden.

With crystals in her bed she was able to breathe better and as I left she suddenly looked at the clock and realized she was four hours late for her pain-killing injection. She was completely peaceful and happy to contemplate using crystal visualizations as a way of being free of the restriction of her body. Cathy had been a dancer, and she realized that she could still dance in her mind and imagination. This peace lasted until she died.

In Cathy's case it was not appropriate to use crystals to work directly on a specific part of the body. On another occasion, in which a girl had a brain tumour removed, I placed three crystals around her head in a triangle. One was put above the top of her head pointing down, the other two were placed on either side of the head pointing in. This triangular crystal arrangement is extremely powerful and brings in the Father-Mother-Holy Spirit aspects of God. It can be placed on or around any area of the body that needs healing. Breathe love and healing into the crystal first.

Crystal healing arrangements

Another crystal treatment for a wound or a pain is to direct two crystals pointing at each other with the wound in the middle. I have used this successfully for broken arms, legs, tennis elbow, sprained wrists and ankles.

One fifteen-year-old, with a sprained wrist, was due to take final school exams and did not think she would be able to hold the pen. With crystal treatment she not only wrote her exam papers, but memorized the subjects better when swotting.

When treating with placement of crystals and gems, the patient must always lie down, unless you are working on a foot. In this case, the crystals can be placed on either side of the foot.

If you need to anchor and stabilize yourself, or someone else, place a crystal, point down towards the foot, on each thigh for a minimum of twenty minutes. If you need to open up, mentally or spiritually, place a crystal in the centre of the forehead, pointing up towards the crown.

Remember that amethyst, sodalite, lapis lazuli, sugalite and sapphire are all good gems to use in this area. The blue stones calm, soothe, clarify thought and bring wisdom. Sapphire alleviates depression. Amethyst, sugalite and sodalite stimulate intuition.

If you feel inadequate and unloved, watermelon tourmaline,

137

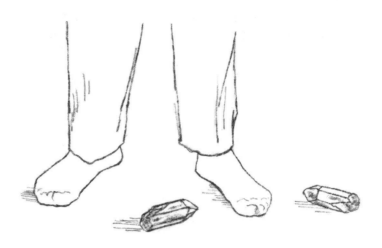

*Using crystals to heal a wound — two crystals are pointed at each other
with the wound in the centre*

ruby or rose quartz are the stones to lay on the heart area.
Green stones will also soothe, heal and bring heart feelings into
balance.

For digestive problems, think of the more orange-yellow
stones such as agate, carnelian, jasper, amber, topaz and
citrine. Remember that citrine is a wonderful communication
crystal and that topaz will stop your energy being drained by
others.

For menstrual pain, rose quartz laid on the stomach is very
healing. I used rose quartz on a friend's caesarian scar. In
addition to the operation she had lost the baby and was holding
on to a lot of physical and emotional pain in that area. So that
she could use the quartz whenever she wanted to, I left it with
her. She told me later that she simply sat with it against her
stomach as if she held the baby that died. She felt as if the stone
was like a sponge that absorbed her pain. In a relatively short
time she recovered and started lending the quartz to other
people to cure their ills and fears. Since then this rose quartz
has been used in many different situations.

For a general physical recharge and to stimulate the circulation, red stones are the most appropriate. Agate and coral

help physical endurance and overall well-being. Rose quartz, amber and kunzite are good for the skin and tissue rejuvenation. Red stones will also clear sexual difficulties.

Using a crystal to 'open up' mentally or spiritually — a crystal is pointed at the crown or Third Eye

Crystals and visualization

In my work I have had to do a lot of public speaking. It started when attending other people's lectures and I suddenly found myself being asked to speak. My terror when this happened was so great that I frequently tripped when getting on to the stage, and fell flat on my face, or caused chaos by dropping the microphone, or, worst of all, opened my mouth and either absolute gibberish, or no sound at all, came out. My hands would go white with fright, I went backwards and forwards to the bathroom and soon became a miserable wreck.

I had earlier noticed that when a speaker gave a lecture, a turquoise colour often emanated from around the throat. A friend gave me a turquoise scarf and I began to wear it constantly, night and day, I imagined the colour swirling through every part of my neck and throat. A little later I bought a small piece of turquoise that was shaped like Sri Lanka, where I was born. It seemed a good omen. I was also given a turquoise bead necklace from New Mexico, which I also wore all the time.

In conjunction with this I used visualization. I closed my eyes and imagined myself first making a complete fool of myself, as I had done before. I then mentally drew big black crosses over this image, and imagined myself successfully giving a talk. I spent much longer on the second image than on the first. I also put notes that I needed to memorize under my pillow at night, with a crystal on top, to help me retain the information.

When I next had to speak I carried a crystal in my right pocket and wore the turquoise around my neck. Although I am still a little uncomfortable just before I start, I can now talk without the paralyzing fear I experienced before. A major phobia has been healed. Turquoise loosens the throat and enables us to say what we have to say — frees us to speak our own truth.

When you use visualization in this way you lower your brain-wave rhythm and reach into the deeper, buried-memory part of your mind. This helps to erase the negative fear and reaction, as you put the thought of what you want to accomplish into your computer brain — a little like programming your crystal — it stimulates a larger part of your mind to co-operate with you rather than fight against you. Holding a crystal in your left hand activates the brain's receptivity to this.

Since then I have used this process to spring-clean a house, lose weight, fulfil a task I am resisting. During the day we act mainly from the Beta part of the brain. When we relax, watch television, begin to go to sleep or daydream, we lower this rhythm into Alpha and when we are asleep and dreaming we lower it to Theta. Delta is the deepest unconscious rhythm — such as that experienced in anaesthetized sleep. If we decide to do something purely from the Beta side of the brain, which is approximately 10 per cent of the brain, then 90 per cent of the brain is like a submerged iceberg fighting against us. If we get the co-operation of even some of the 90 per cent we draw on a strong, supportive energy.

A girlfriend, who had tried every diet imaginable, used this technique to reduce herself from 18 stone to 11 stone. In addition she put a programmed crystal next to her plate and at every meal said 'Everything I put into my mouth is going to make me slim and beautiful.' She stopped dieting but found that she lost weight anyway. Six months later I almost did not recognize her. She had turned herself into a new woman.

Many years ago I became a vegetarian. From one day to another I stopped eating anything that was not raw fruit or vegetable. I felt marvellous. I was full of zest and did a job that required me to drive thousands of miles and deal with hundreds of human, personnel problems while marketing fashion through many retail outlets. Twelve months later my hair began to fall out — not only my hair but my eyebrows and eyelashes fell out too. I panicked and went to a doctor who told me I was crazy to become a vegetarian overnight. He explained to me how the body functioned and that if I wanted to change diet it should be done carefully and slowly. He told me that I must eat protein in the form of fish or I would become ill.

He said, 'I will prescribe fish for you as if it were medicine, you must eat it.' I did not want to be bald and so I went to a fish shop. I waited in a queue and at the very moment that it was my turn to be served I saw what I thought was a dead lobster walk. I was revolted and walked away. I decided that fish were really another form of seaweed and that my body would benefit from kelp (dried seaweed) as much as it would from fish. I tried to swallow large spoonfuls of powdered kelp and nearly choked to death. I then remembered all the miraculous things that had happened when I used my mind and imagination and crystals. I followed the same formula which had enabled my friend to lose weight.

I empowered (empowering breath as described earlier) two or three crystals with the thought of my hair being thick and healthy. I placed them near me when eating, sleeping or working. With every mouthful of my live, raw vegetarian food I said, 'This is going to make my hair grow strong, shiny and thick.' It worked. Like healing my face it did not happen overnight but the hair on my head grew back. It is now so thick that hairdressers complain when it has to be blow-dried. (My eyebrows never grew back more than a vague line but then I really did not care enough about them!)

Balancing the four elements

If you are a water person who flows with feeling, an emotional person, aquamarine and amethyst can stop you drowning in your emotions or flooding other people with them.

If an earth person, someone who enjoys the physical things of life, a smoky quartz will balance you if you tend to be aggressive and dominating. Rose quartz will also soften you.

As an air person, someone who is inventive, mentally stimulating, living from wits and intellect, citrine is the gem to ensure the expressing of your ideas, should you become too locked in your thinking.

For the fire person, who is someone who can inspire others with the power of rhetoric but who also tends to have low self-esteem, clear quartz can prevent taking in criticism.

If there are difficulties in relationships, the fire person will withdraw, sulk, and, when everyone else has forgotten all about it, suddenly snap out with a punitive verbal attack. The water person will cry and feel 'How could you do this to me?' The earth person is more likely to physically lash out and punch a nose or slap a face. The air person will cut off all feeling and mentally assess the situation in a cold, clinical way.

The stones I mention balance the positive and negative in you. To stimulate other qualities you need to use stones with different characteristics. For example, citrine will help to stimulate the water person's mind so that he can act rather than react, communicate from mind as well as feeling.

A wonderful healer and friend, Ripley, who lives in South Africa, told me that she treats toothache by getting the patient to hold a crystal under the tongue. She said that even with chronic toothache the pain goes within half an hour. Ripley is very attuned to her crystals and she often asks them what they want her to do. Once they told her to place a patient in a crystal triad, with a female crystal below the left foot and a male crystal below the right foot, both pointing in, and an amethyst on the forehead. After about 9–15 minutes she picked up another crystal and made a figure eight movement around the body, followed by rapid circular motions into the heart and down to the solar plexus. The patient suddenly burst into tears and emerged completely healed.

Ripley was also visited by a Zulu Sangoma – an African medicine woman. This woman said that she could no longer hear her wisdom voice, the voice that told her how to make her magic potions. Her physical hearing was also muffled. Ripley suddenly sensed that she should put a crystal in the Sangoma's ear. After three or four minutes the woman exclaimed 'I can hear, I can hear!' and from that moment she too was cured.

Absent healing

Ripley uses crystals every day to broadcast or transmit healing to people who have asked for it. To do this, you can either put the person's name or photograph under the crystal, or you can impregnate, with the breath, the thought of the person into the crystal. Once you have put someone inside a crystal, the crystal will continue to transmit to them. When you put the crystal on top of the name, the touch of your hand combined with your intent will automatically programme it. If you wish you may also programme the crystal to send love and vitality to anyone whose name is placed under it.

This type of broadcasting or absent healing works no matter how many thousands of miles away you might be. It is similar to tuning in to a radio programme broadcast from a country on the other side of the world. The Russians tested the power of thought transmittance, and, in one experiment, took baby rabbits away from their mother and placed them in a submarine many miles away. They monitored the mother's reaction while they killed each rabbit, and, at the exact moment that each one died, the wire attached to her body jumped. Love and caring create a connecting link to which there is no barrier.

Healing methods

There are two ways of using crystals. One is by putting them in a static role by placing them in certain configurations on or around the body. The other is to use them in a mobile role, by holding them and making clockwise or anti-clockwise movements over a body.

When using them like this there are one or two things to remember. The left hand is the receptive hand, the right hand is the transmitter hand. To pull pain out use the left hand, to direct healing in, use the right hand. A counter or anti-clockwise movement draws out pain and disperses it. A clockwise movement focuses and directs the energy.

When holding a crystal for body work, it is important to clasp it with the base pressed firmly against the palm (see the picture on page 95.) You must ensure solid connection with the female, solid or opaque or flatter end, and direct the pointed end at the body. Do not let the crystal dangle loosely from your fingers. Use crystals that are big enough not to be hidden by your hand, but not so big that they are unwieldy.

To draw pain away from the body, hold the crystal three to four inches above the spot. Move it slowly in an anti-clockwise circle. Think of the crystal as a magnet drawing the pain away. Continue as long as you feel or intuit is enough. Or count three times, seven, nine, even up to twenty-one times. Always move the crystal anti-clockwise.

You may not be able to draw all the pain out at once. Like taking medicine, you often need more than one dose. When you feel you should stop, stop. The more aligned you are with your crystal, the more you will sense the exact moment to do so.

Remember that the crystal balances and transforms. What it takes in, it also puts out, changed. However, to clear the crystal of any possible residue, shake it two or three times, or breathe a cleansing breath into it. Send to the light anything it may have taken in. In order to direct healing energy into the body, now that the pain has been extracted, rotate the crystal clockwise over the same area, timing it in the same way as you did in the anti-clockwise movement. Finally, hold the crystal absolutely still over the same point, visualizing light and healing pouring through you and the crystal into the patient.

If your hand gets tired, put the crystal down for a moment or two, stretch your fingers. Relax. You may find that you as well as your patients feel sensations of coolness, heat or warmth. Whatever you experience, enjoy it. I find I receive as much benefit as my patient.

When I began learning about crystals, I had many dreams

illustrating how to use them. I saw healing temples where robed figures used their hands to scan the aura or energy field of a person who appeared to be malformed. Crystals were then set in a pattern around the person, and with the use of light, crystal energy was directed towards them. Some of the robed figures worked directly on the aura, using crystals. Others stood in front of giant crystals – eight feet high – and seemed to draw crystal energy into their solar plexus and then redirect it to people who were sick.

In the dreams, it looked as if the healers were able to sense with their hands the pattern of health and perfection for each person. They then used crystals to charge in each person the memory of this perfection. I saw this work focused more on the aura than the actual body, and that as the aura became whole, so did the structure of the body change and become whole too. *(See note at end of chapter.)*

Auras

Crystals are exceptionally effective. Not only is their composition like that of our own bodies, which is why our own bodies respond so well to them, but they also deeply penetrate the electro-magnetic field around us as well as our subconscious. This electro-magnetic field or aura is made up of a number of subtle or invisible bodies which swirl and move, expand when we feel happy, shrink when we feel depressed.

Most auras have a mixture of colour, although one usually predominates. It is like a bubble or an envelope and contains within it the pattern of who we are, the non-resolved problems we need to deal with, which show up as static interruptions of the electro-magnetic field. When someone is sick or in trouble, they put out a distress signal in the aura, which the crystal tunes into and acts on.

Wilhelm Reich described the armouring of emotionally victimized humans. Their muscles lose the ability to relax, their breathing becomes shallow and there is a marked loss of vitality. All this shows in the aura. Chronic drug and alcohol addiction can affect the aura so that it becomes worn and sometimes develops holes. The aura is our very own protective space. When people infringe it, such as by pushing a supermarket trolley into us, and if we are normally sensitive, we will feel extremely uncomfortable.

Illness shows up in the aura before it manifests in the body. Learning to see and read the aura, so that action could be taken before the body was affected, would be good preventative medicine. Accidents show up in the aura three to seven days before they happen. In other words, there are no accidents. In New York in the 1920s and '30s, Dr Harold Saxon Burr conducted hundreds of experiments measuring this area around the body. He discovered that he could predict illness before it happened and went on to discover that he could predict accidents too. Later still he found that he could predict when a woman would become pregnant because her aura expanded to that of two people six weeks before conception took place.

When imbalance shows up in the aura, the static inter-ruption of the electro-magnetic energy looks like a shadowy blob or discoloration. Crystal energy breaks up and disperses this while opening our subconscious mind to deal with it. This is why crystals can sometimes cause discomfort to the patient while being used — arousing sudden emotion, such as tears or rage — as part of the healing process.

Some stones are meant to be lifetime companions. Others have power for a particular situation and then need to move on. They will often disappear when their work is finished. You can always communicate with a crystal telepathically — think of it with love, bless it, thank it, release it. This is easier to say than do. I was staying in a house in Los Angeles when my favourite crystal suddenly announced that it wanted to stay with my hostess. I asked all my other crystals, brought to America for a workshop, if one of them would remain instead. They all said 'no'. It took me three days of asking different crystals over and over again. 'Isn't it you that is meant to stay — or you — or you?', and always getting the same answer before I was able to pass it on. I breathed love into it, wrapped it up and presented it. A week later, in a different country, I received another, almost identical crystal. I learned that my refusal to let go blocked what was trying to come in. I also saw very clearly that when one door in life closes we should turn around and find the door that is open instead of continuing to gaze at the door that has closed.

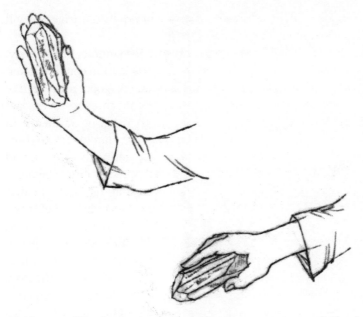

Hold a crystal like this for the energy field/aura massage — turn the hand
lengthways to do the massage

The aura ought to be cleansed and cleaned as much as we
brush and clean our teeth. It can be done with the hands in a
brushing downward movement around the body from the top
of the head, but it is much more potent if done while holding
a crystal lengthwise. When you have done this a few times, you
will notice a dramatic difference both physically and mentally.
Brush down to disperse anything you do not want — remember
the aura extends beyond the body so keep your crystal about
6–12 inches away from the body. When you have finished
brush upwards from the feet to recharge the aura. I do this
myself two or three times a day and it centres and refocuses me
if I am becoming unclear. You may find it works better for you
to get someone else to do it to you. Children love cleaning
their auras.

The Chakras

Other areas that can develop problems by becoming blocked
are the chakras. The chakras are seven vital centres that go

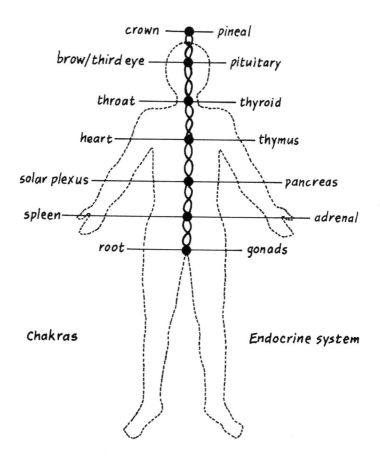

crown —— pineal
brow/third eye —— pituitary
throat —— thyroid
heart —— thymus
solar plexus —— pancreas
spleen —— adrenal
root —— gonads

Chakras   Endocrine system

from the top of the head to the sacrum at the base of the spine. They are like power points where lines of energy meet and cross. Chakra in Sanskrit means wheel and each chakra when healthy is a whirling vortex of energy. There are many smaller chakras but the main ones are the Root, Spleen or Sacral, Solar Plexus, Heart, Thyroid, Brow or Third Eye, and Crown. These seven vital centres are connected to the spine by cords. We talk about the flowering of the chakras because as they develop they look like flowers. The chakras control the organs and glands of the body and each one relates to a different frequency on the colour scale.

The Root chakra is connected to the base of the spine and is associated with the reproductive organs. The related colour is red and the gland is the gonads.

The Spleen chakra is just below the navel and is the area where we feel 'gut-level' emotion. The related colour is orange and the gland is the adrenals.

The Solar Plexus chakra is at the diaphragm and is a chakra of 'taking in and giving out' energies in relation to others. If you have a strong sense of self it will be strong, if a poor self-image, this area will be weak. The Solar Plexus is where we 'tune-in' and psychics often develop a paunch there to protect it. The related colour is yellow and the gland is the pancreas.

The Heart chakra is at heart or thymus level. It is the chakra of love and the midpoint between heaven and earth, time and space — and a very important seat of energy and wisdom in the body. The related colour is green and the gland is the thymus.

The Throat chakra is at the throat and is the seat of communication. The related colour is blue and the gland is the thyroid.

The Brow or Third Eye chakra is at the centre of the forehead between and slightly above the eyebrows. It is the seat of spiritual intuition. The related colour is indigo and the gland is the pituitary.

The Crown chakra is located at the top of the head and is the entry point for various energies flowing through the body. It is the area of spiritual opening. The related colour is violet/purple and the gland is the pineal.

The flow of energy between the chakras affects our physical health as well as our sense of well-being. When working on the chakras with crystals it is important to go with the natural clockwise, or anti-clockwise movement of the chakras. If you make an anti-clockwise movement on a chakra whose natural flow is clockwise it is a little like taking a scab off a wound, it is far too rough and can cause discomfort.

In a woman the chakra is anti-clockwise at the crown, clockwise at the brow, anti-clockwise at the throat, clockwise at the heart, anti-clockwise at the solar plexus, clockwise at the spleen and anti-clockwise at the root. In the man it is the exact opposite. This is why men and women literally turn each other on or spark off each other. A negative, or anti-clockwise vortex or spiralling energy hits a positive or clockwise vortex and it stimulates a reaction similar to charging a battery with an electric current.

The etheric meridians

In addition to the aura and the chakras we have etheric meridians. A meridian is a subtle line on which energy flows, like a wire, carrying and circulating the life force into and within the physical body. To get a sense of the etheric meridians, imagine your etheric − or more subtle, invisible − body as a larger, shadowy outline of your own body. The etheric meridian lines of energy run from the top of the head of this body to the points of the fingers and toes.

When the energy flow is blocked in the chakras or the meridians, or when the aura is showing imbalance, it will always affect the physical body. Acupuncture is one method used to stimulate the movement of life force in the body. It is also a wonderful preventative therapy. In China where it originated hundreds of years ago you did not pay the acupuncturist if you became ill. It was his function to keep you well and healthy.

Cleansing the body's energy lines

Another way of clearing the aura, the chakras and the etheric meridians is to do it with crystals. To balance your own chakras lie down and place a crystal, point up towards crown, on each chakra for about ten minutes. This can be done twice a week. It will recharge the cells of your body as well as your auric field.

If you have only one crystal start with the base or root chakra, move the crystal up the body − allowing approximately three minutes on each chakra − and finish with the crown.

To place a crystal on an area of discomfort in your body is another excellent treatment to unblock both your body and your aura. With the crystal in place think of the pain and any emotion or a memory associated with it. This may come into your mind as a clear understanding of the cause of the problem, it may appear as a symbol or a colour. Focus on the image and then exhale sharply, two or three times. Sometimes this will catalyse other memories which, as they too are released, will facilitate the healing process.

I used this method with a young woman who had growths in

Self-work on the Crown Chakra

her uterus. During the session I asked her what the growths symbolized. She replied, 'The child in me that wants to be born but I deny it, the mother in me that wants to give birth but I do not allow it.'

Her own reply gave her such a shock that she sat up abruptly. We were then able to discuss how she became pregnant when she was seventeen. In panic and despair she had the baby and gave it away. At the time she felt that she had no choice. She also denied feeling regret or guilt. However within three years she developed growths in her womb which had to be operated on. In the process some of her womb was also removed.

She was now twenty-seven years old and happily married to a husband who wanted children. This was possible but only if she was extremely careful. She had been told by her doctor that, if pregnant, she would have to spend most of the nine months lying down. She was not willing to do this and then the

Self-work — 'witness' on the Thymus

growths came back. We communicated to the growths in her body and worked out a healing formula which she could use on her own. This combined the use of crystals with affirmation, visualization and meditation. It took about eighteen months of concentrated effort but she was able to heal herself without surgery. In her early thirties she gave birth to a beautiful baby girl.

When stimulated by crystal energy we can more easily identify the nature of a problem and see how to deal with it.

Another crystal treatment that allows the healer to communicate with another person through the subtle bodies, and frequently activates a dramatic release of blocked energy, is the one used by Marcel Vogel. This involves the Witness or Thymus area, so called because many religions believe that it is from here that we witness our being. Hold the crystal in your right hand, charge it with your breath and rotate it counterclockwise, three or four inches away from the thymus of your patient. Continue until you feel you have made a connection and then let the point of the crystal touch the body. The left hand should be held, palm open, a few inches away from the back, between the shoulder blades. This puts you, as a healer, deeply in touch with the inner essence of a person and can catalyse instant release. No matter what form the release takes — tears, coughing, a sudden physical movement — always have the person exhale sharply two or three times as well. Continue to hold the crystal at the thymus and visualize white light pouring into the area for a few moments afterwards.

When I use this technique I frequently feel energy, like a coiling snake, pouring into my left hand and arm. I pull it out and throw it to the light to be transformed. I did this once on a beach and closed my eyes while pulling the snake-like coil of energy out of my patient. When I opened my eyes I saw that we were surrounded by a circle of birds who were absolutely still, silent, attentively watching. It was a magical moment and the young man's asthma never returned. His life had been traumatized by an exceptionally difficult relationship with his father. During the session he understood the nature of the problem and how to sort it out. This is what enabled complete healing to take place.

Aside from aligning to the witness area of a patient I sometimes use a crystal, held lengthwise in my right hand, to scan the body. I let the crystal guide me to any imbalance. This may be sensed as warmth or cold or just as an inexplicable 'difference' in the electro-magnetic field. I then rotate the crystal, point down, in an anti-clockwise motion, using my left hand behind the spot to receive or pull out the energy as it is

Healer using a crystal to scan a body during a session

released. I then rotate the crystal clockwise to direct healing into the area until it feels appropriate to stop. Before removing the crystal I visualize white light pouring through it and into the patient.

All these methods are simple and effective. It is important to try them for yourself and find out what works best for you and your patient.

Another technique, which is the way in which I complete all my own healing sessions, involves working directly on the chakras, the aura and the etheric meridians, before sealing the aura. Emotional release, which is often part of the healing process, puts a static disturbance in the auric field and sometimes clogs one or two of the chakras. This method clears away any residue from the session and allows the patients to go home in a crystal-clear state. It can be used after any other therapy as well as being a treatment on its own.

Creating a healing environment

Before you start it is important that you create a peaceful atmosphere. The room should be clear and light. Crystals, candles, fresh flowers and harmonious colours reflect spirit and disperse negativity. If you have a room that you can dedicate as a healing sanctuary, a wonderful energy builds up. Because I work all over the world I am unable to do this but I always choose to work in rooms where the colours are light and bright, places where a patient can come in and immediately feel comfortable and relaxed.

I try to create an atmosphere in which a person feels enfolded in love and peace. This means preparing yourself as well as the room. Try and keep your own body as healthy as possible. Take time to be still before your patients arrives. Do not frantically puff the cushions, or scream at your child who is late for school, as the front door bell rings. This sort of frenzy can be felt as soon as you step into a house.

An African Sangoma, or medicine woman, says,

> This healing work demands that anyone entering
> your house should be able to see that your face is

serene, your heart is clear and that you are welcome. The way you make them welcome is therapeutic for them. Should a person enter my house and discover that I am unhappy or angry about something, this will have a detrimental effect and will not aid his recovery.

When people arrive they are sometimes a little apprehensive and need reassurance about what is going to happen. Although it is not appropriate to play music throughout my three-hour sessions, I do play a meditational-type tape at the beginning as an aid to relaxation. It is essential that the patient be as relaxed as possible. If his or her body is tight and stiff it tends to reject the healing.

Before I begin I get the patient to lie down and I place crystals around him or her. I usually put one on each corner of the bed pointing inwards. Sometimes I use five crystals and make the shape of the five-pointed star around the patient. This is the symbol of the perfected man and creates a powerful framework within which to work.

I then check the chakras. This can be done by hand scanning — using your hands just above each chakra to sense the flow — or with a crystal or with a pendulum. I do this in order to compare the before and after of my treatment. After the session I recheck the chakras.

I always say a prayer asking that only what is appropriate for the highest good of the patient may take place. I breathe with the patient while asking him to breathe in light until he becomes relaxed. This allows our energies to blend.

I attune to the crystal by rubbing it, holding it and sometimes by rotating it in a clockwise movement above and into my left palm until I sense that the connection is made.

I work on the back first so the patient must now turn on to his stomach. I hold the crystal firmly in my right hand, point down, and, starting from about twelve inches away from the crown, I make big clockwise circles all over the body coming down the left side and up the right. I am working on the etheric body so I cover a space that extends about 12 inches away from but includes the physical body. I do the entire area three complete times.

I then hold the crystal lengthwise and brush the energy up the body from toe to crown, before turning the patient on to his or her back. I spend much more time working on the front of the body than the back.

With the crystal still held in my right hand, and working from the crown to the root point, I circle the crystal 2—3 inches above the chakra in anti-clockwise or clockwise movements, depending on whether it is a man or a woman (see page 149). This clears any residue of the session from the chakras.

I sense when each chakra has had enough but if you do not feel you can trust this, count 3 or 7 or 9 times. I breathe-cleanse the crystal after each circular movement of each chakra. I then hold the crystal over the chakra for a few moments allowing light to flow through before moving to the next chakra. The first movement is a little like using your finger to clear a blocked sink. The second is like letting the water pour through freely.

Having done that I move my crystal to behind the top of the head and I make big clockwise circles all over the etheric body in exactly the same way as I did to the back. Again I cover the entire area three complete times, coming down the left side and up the right.

I then go to the top of the head and run the crystal down the etheric meridians, following the lines to extend beyond each finger and each toe. This means you run the crystal down the body ten times. Use a firm brushing away movement. I start with the left side and move to the right but it is not important on which side you start.

I then hold the crystal lengthwise and brush the energy up the body from toe to crown. This recharges the aura and the muscles.

I finally seal the aura by holding the crystal point down, and starting from the solar plexus, make two figure eights that encompass the whole body, i.e. one circle of the eight encompasses the upper body, one circle encompasses the lower body. I bring the point of the crystal back to the solar plexus and say words to the effect of 'I seal this aura in the love and the light of God'. As I pull the crystal away I place my hand over the solar plexus for a few moments.

Always let the crystal energy soak in afterwards. Do not let the patient get up straight away as he may feel a little light-headed or 'spacey'. Breathe cleanse your crystals before your next patient.

It sometimes takes two or three weeks for the full benefit to be felt and during that time it is possible the patient may feel a little uncomfortable. You have cleared away some of his or her crystallization and this can make him a bit wobbly.

Quartz crystal, because it contains within it the qualities of every other colour, can be used successfully on all the chakras. It will recharge, stimulate and clear away any blockage or stagnation.

Healing with coloured stones

If you prefer, you may also use the colour of gemstone that relates to the specific chakra. For example, a red stone when placed on or just above the genital area will affect the root or base chakra. It will regenerate the cells and the circulation, renew energy and vitality. The carnelian, agate, bloodstone and garnet are all suitable gems for this area. If you use a clear quartz crystal in conjunction with another gem placed on each chakra it will add to its effectiveness.

For the spleen or sacral chakra just below the navel, orange or yellow stones such as citrine, amber or agate or rutilated quartz are the most beneficial. Golden or brown topaz, citrine, amber that is clear rather than opaque, will activate the solar plexus. Placed on the solar plexus, one or other of these stones stimulates giving and receiving, communication, attunement and sensitivity to others. For the heart chakra and thymus, the green and pink stones are the best. Malachite, emerald, moss agate, rose quartz, kunzite, plain green or watermelon tourmaline will balance the organs and systems of the body, while enabling you to love yourself and others.

The throat chakra will respond to blue gems. Aquamarine, turquoise, lapis lazuli or sapphire placed on the throat will soothe sore throats and assist in speaking one's own truth clearly. Sodalite, sugalite and amethyst are ideal for the Third Eye chakra in the centre of the forehead. Clear crystal or amethyst placed at the crown will enable communication with one's own inner wisdom or intuition. It can be a powerful experience to lie down with a different gemstone on each chakra. Remember that the colour of the stone, even if not a precious gem, can have the same effect as a jewel. The more dense or opaque the stone, the more physically balancing it is; the more clear or brilliant the stone the more spiritually uplifting it will be.

Although I believe it is beneficial to get somebody else to work on my chakras if they are going to rotate the stone rather than place them on my body, I have used my small Herkimer diamond crystal with great success on my own chakras. I hold it in my right hand and rotate it anti-clockwise alternating to the root and have found myself in a deep state of relaxation afterwards.

Body symbolism

Like crystals the body, which is itself a kind of crystal, is far more complex than is generally thought. To interpret body symbols, in other words to learn to read the message being given by the body when it breaks down, is another way of gaining understanding of both your life and your subconscious attitudes to life.

To get to know an artist we study the artist's paintings. They are the expression of his or her vision and creativity. When we examine them we begin to sense something of the artist's personality — we get a feeling about him/her.

In the same way we need to study our bodies. They too are the outer expression of our own inner vision and creativity. The more we examine and communicate with our body the more we can understand our subconscious patterns.

I began to discover the importance of this when, frequently, I became ill, accident-prone and hospitalized. If I could find and heal the root cause of a problem the physical symptom would usually disappear. This meant that I had to apply what I learned, act on it in some way, which invariably led to a change of attitude. For example I used to suffer from migraine. I discovered that problems with the head, whether migraine, brain tumour, mild headache, or even a knock on the head, are usually caused by conflict with authority. The authority may be that of boss, husband, wife, father, mother, the policeman who gives you a parking ticket, the head of an organization, it may also be resistance to your own authority, refusal to take responsibility for some aspect of your life. In my case I constantly gave away my power, apologized for everything I said and did not stand up for my beliefs. As soon as I recognized this and began to live my life in a different way the migraine attacks stopped. They have never returned.

I also saw this happen with other people. A woman on the verge of open heart surgery as a result of three major heart attacks, asked her body during a meditation, 'Why are you doing this? What are you trying to tell me?' The immediate response was, 'Your heart is trying to open. You must develop unconditional love for all those around you, not just your husband and family.' Her husband was building an old people's home. She believed her responsibility was to support him rather than the old people. This answer made her think about her attitude and she realized that her first heart attack came soon after her husband asked her to participate and she refused.

She suddenly understood that she was afraid of being so busy with the old people that her family would be deprived of proper care. However her children were now grown-up and no longer needed her care. This realization did not prevent the open heart surgery but prior to the operation she spent time talking to her heart. She said, 'Thank you for the message. I understand it. I will act on it and you do not need to give me the same message any longer. You can now become strong and healthy again.' During the operation a group of friends surrounded her with light, love and crystal energy, which helped her body to heal more quickly. Ten days later, she was out of hospital and planning her future activities within the old people's home. She is now completely better and leads an active and enjoyable life.

Our bodies in sickness will give us the message we cannot, or do not want to hear, when we are healthy, on any other level. There are hidden channels of communication from the subconscious and unconscious to the conscious mind.

If you relax, close your eyes and ask the questions you need answered, these channels can be tapped. Practice is the key, and attunement to the highest within you – your inner wisdom or God self – is your protection against irrelevant answers. Sometimes, it is easier to do this if someone else asks the questions. When you question your body speak directly to the organ or limb that is sick as if you were questioning a person. 'Why are you doing this? What do you want me to know or learn? How can I heal you?' are the type of questions to ask. Continue the conversation until you are satisfied. Remember that a crystal held in your left hand will help you to receive the answer more clearly.

When you question the symptom to find the cause remember that your body is not punishing you. You have subconsciously programmed it to behave exactly as it is behaving. Many of us become ill when our bodies need a rest, or we feel unloved and don't know how else to get attention.

Much of this begins in childhood. We pick up babies who cry and ignore them when they gurgle and chew their toes. We give more attention to children who misbehave than we do to children playing happily. It is natural to do this but it also sows seeds in our subconscious that we matter more when we are suffering and in pain than when we are happy. As an adult this can bear fruit by our becoming sick when we consciously or unconsciously feel neglected or ignored. If we give more encouragment to joy and happiness in childhood, and less to misbehaviour, it would sow very different seeds for the future and we would be far happier.

In working with your own body when you interpret the symbol, or hidden meaning, of a broken arm or a chronic skin complaint you need to examine both the inner and the outer message to find what is true for you. For example thyroid problems are associated with being over-critical. More often than not it is criticism of self rather than other people, and usually stems from lack of self worth.

If you have heart problems, low blood pressure, high blood pressure, these are all tied in with love – inability to give, or maybe receive love in the case of low blood pressure, loving too freely, too rashly in the case of high blood pressure. Your inability to give or receive love freely may come from deep hurt in your childhood and does not mean to imply that you are nasty or selfish.

The structure of your body represents the structure of your life. The spine stands for will, human will aligned with divine will. Problems with your spine often indicate conflict with your own inner wisdom or divine self. For example, perhaps I sense that I have a gift to heal but I am too afraid to leave the security of a well paid job to actually use it. This is a conflict of which I may be well aware or it can be quite hidden and only manifests itself in an aching back. Curvature of the spine usually occurs when there are constant battles between a growing child and a domineering mother. The child's spine says, 'All right, I give in, I surrender.' Lower back pain is your subconscious saying, 'It's too much, I can't cope. I am overstressed.' Medical research in America shows that patients who smoke are twice as likely to have acute back problems as those who do not.

Your arms signify implementing the ideas of your life. We will frequently break an arm when we either block ourselves, or are held back by an outside situation or influence. The arms of a man I know, whose life work was suddenly stopped, became so badly affected by nettlerash that for a week he had to sit, immobilized, with his arms in a basin of liquid. During that time he regained the courage to look for something else to do with his life.

Legs mean support and if we feel a lack of support from others, or insecure in ourselves, we will often damage a leg. This lack of support may be mental, emotional, spiritual, physical or financial.

Your ankles symbolize turning points towards a new direction in life. If you are uneasy about this new direction you might trip and twist your ankle. Problems with your feet often indicate fear or insecurity in stepping fully into life. The feet are one of the most sacred parts of the body, and they connect us to the earth. To bathe or massage someone else's feet signifies a willingness to care for them, rather as Jesus washed the disciples' feet. To soak your feet in crystallized water, water in which you've placed a crystal, before you go to bed at night can be a wonderful preparation for sleep. Soak your feet for five to ten minutes and imagine all the events of the day floating away into the water. You will sleep better and your feet will benefit.

Your neck and joints are to do with flexibility and we literally say that someone is stiff-necked when we mean they are

161

inflexible. Pain in your knee joints is connected to pride. It is stating that you need to kneel, be humble, surrender an idea or opinion you may have about yourself. This pride can also be that of keeping a stiff upper lip in difficult circumstances. Joint pain, such as with arthritis, is caused by too much acidity. Too much acid means holding on to bitterness, resentment, inability to forgive.

Your face is the expression of your personality. It is the first part of yourself that you show to the world. Your eyes mean the ability to see clearly. Short sight signifies being preoccupied with the details, you can't see the wood for the trees and are a bit of a perfectionist, while long sight implies a detachment from what is under your nose and the ability to envision and plan far ahead.

Blindness often stems from a refusal to see or carry out one's purpose in life. It can also be a karmic debt from another life. I worked with a man who had gone blind and he suddenly burst out, 'I cannot bear to see what is happening in the world. It is all too terrible.' Another blind man refused to recognize and use his psychic powers of clairvoyance with which he could have helped others. Sometimes blindness will occur to push us into developing our inner senses.

Many people change completely after acute physical problems push them into a reassessment of their lives. Dis-ease is a bio-feedback system that says, 'look — listen'. Illness is an opportunity to change and we should try and accept it like that.

Ear problems and deafness usually reflect a refusal to hear. This may be a conscious decision to 'turn a deaf ear', when you are bored, provoked or being nagged or an unconscious cutting off to the outside world due to some traumatic event. People with ear problems are often impatient to do what they want to do when they want to do it, they do not wish to listen or be receptive to others' needs or opinions. Many children suffer from hearing difficulties today. They are souls born into a new era, they have a different perception and do not want to hear all the old stories.

Your mouth is the feature through which you speak your own truth. It is part of your face and personality and is, like your throat and voice, part of your identity. Chronic problems with teeth stem from neglect of self — physically and emotionally.

Shoulders symbolize responsibility. Pain or knots in the shoulders suggest that you feel burdened with too many responsibilities, real or imaginary.

Your lungs are associated with emotion. I find it interesting that men tend to get lung cancer more than women when, for a long time, men have been encouraged to repress their emotions. Women get more cancer of their reproductive organs than men do of theirs. Since Victorian times women have been taught to deny their bodies, to believe that they are somehow unclean, something to feel slightly ashamed of.

A woman's breasts symbolize the outer expression of femininity while the womb represents the seat of more profound female power. Heavy breasts denote maternalism which can become over possessive. Breasts or a thick, broad chest in a man mean the ability to protect, verging on over-protectiveness.

A man's reproductive organs are linked with his capacity to express himself freely in life and work. Men who have stifled themselves or felt blocked from true fulfilment will often develop problems here.

All the organs of digestion are linked with an ability to assimilate or release not only our food but everything that comes into our lives. This may be anything from people, events, ideas to possessions or work or information. If we mis-construe the people around us we will often have liver problems, if we are too self-indulgent the pancreas will stop working properly, inability to handle stress can affect the gall-bladder, repressing emotion, refusal to cry affects the kidneys. Repressing emotion will also lead to nasal ailments like catarrh, sinusitis and asthmatic problems. A cold is also often our body's way of releasing unshed tears.

Chronic emotional pain can also affect the colon causing colitis — this is also connected to a refusal to let go, as is constipation. I have found that constantly held back anger can also damage the liver as well as every other part of the body.

Our emotions have an energy. The body does not care how that energy is expressed. Walking, running, jumping, laughing, talking, singing, shouting, scrubbing a floor, digging a garden are all wonderful ways of releasing anger and upset. When emotion is constantly suppressed it begins to break down the chemical balance of the body, when stress becomes distress illness occurs.

The hips mean free will and independence. Think of young girls jauntily swinging their hips while in the same way elderly people who are losing their freedom will often break or seriously damage a hip.

Hands and fingers express how you use your energy. Thumbs are associated with will but more personal will than the spine. If I quarrel with someone, which is a battle of wills, I will often bang my thumb or catch it in a door. The index finger represents authority in a slightly bossy, superior way. 'I told you this would happen,' says Aunt Ermintrude, wagging her finger at you. A woman I know, who continuously used her fingers in this way, lost her right index finger while mowing her lawn. Instead of listening to the message she began to use her left index finger instead. Believe it or not the mower took that one as well. This did finally stop her. The middle finger is how you use your creativity in your work. The ring finger denotes sexual energy. In a workshop I noticed at least six men with a damaged ring finger. I asked them about this. In each case there was a problem with wives or girlfriends. The little finger signifies freedom — the freedom to express yourself in any way. The greater the flexibility of your fingers the more free you are to flow with life. General tightness of movement in the joints shows that you tend to hold yourself back.

Skin mirrors self-worth and self-acceptance. Chronic skin problems denote inner feelings of worthlessness. I see AIDS too as symbolic of a deep sense of inadequacy and fear that literally destroys our defence mechanisms. We use the phrase, 'thin-skinned' to imply sensitivity. You can also develop rashes, burns or bites if you are irritated or bugged by what is going on in your life. Even a corn on the toe suggests you are feeling squashed, or hemmed in — and not merely by your shoes!

There are many other organs and systems of the body, each holding a hidden message when they break down. This message may be of a physical, mental, emotional or spiritual nature, although they are all interconnected. If you decide to study and question your own body do it with humour and do not become neurotic every time you sneeze twice. You may like to read the book, *Who's the Matter with Me*, by Alice Steadman to explore the subject further.

Even the language we use can affect our bodies. Doctors in America, doing research with patients in hospital, found there was a direct correlation between the words most frequently used and illness above or below the waist. We programme our subconscious when we repeat statements such as, 'I can't stomach this', 'I won't stand for that', 'You make me sick', 'You'll break my heart', 'Back off', 'You're a pain in the neck', 'I can't see what you mean'.

Our bodies respond to affirmations of health and well-being as well as to negative auto-suggestion. If you are in the habit of using phrases that could affect your health change them. 'Every day in every way I am getting better and better' was the affirmation used by Emile Coué at the beginning of the century when he began experimenting with hypnotism. Our subconscious computers are so responsive that it only takes twenty-one to thirty days of repeating such a statement to have an affect.

You can also programme your crystal to stimulate good health by breathing your affirmation into it and carrying it around with you.

The most healthy thing you can do for your body is to enjoy your life — take time to do what makes you feel happy and fulfilled. Practise treating yourself as your own best friend instead of your own worst enemy. Within the area of personal responsibility to your family, friends and work try and do what makes you feel good rather than solely what satisfies others. Use crystals to stimulate a loving relationship with yourself and all of life.

Having crystals in your life is like being in the presence of a person who stimulates and uplifts you. This person may not say or do anything particularly important but you feel brighter in their presence and often encouraged to take some sort of action that you may not have been clear or confident about before.

Most of us at rock bottom have a very non-loving attitude to ourselves. It is one of the greatest problems besetting humanity today. It is a wound which if not healed prevents us from fully participating in life. Unlike illness, which can catalyse change, poor self-image is like a festering sore which affects everything we do. Part of our poor self-image comes from our denial of the parts of ourselves we do not like or trust. Crystal energy can activate the best in us as well as stimulate exploration of the shadow, hidden sides of ourselves.

Note from page 145

A study done in Italy twelve years ago on a cross-section of children found that 60 per cent of children under the age of six would discuss the energy field or fuzz they saw around the body. In the USA another study showed that 20 per cent of adolescents lost the ability even though they had seen auras as a child.

Rosalyn Bruyere, the American healer, did a programme for the Department of Parks and Recreation teaching adolescents who were in trouble to see auras. After this they had a 50 per cent drop in drug-taking. For most of them drug-taking lost its savour if everyone could tell by looking at their auras, what they had taken and when. Seeing auras can help diagnose learning difficulties in younger children.

Crystals and the New Age

> Listen as you walk upon the earth. Listen to all
> the beings coming to life, coming into their own,
> growing into their new expressive selves. It is the
> time of manifesting the dreams that were brought
> to us during the time of the North, the winter. As
> you look around you will see that it is a time of
> love. A time to watch all creatures doing their
> dance of life. (*Painted Arrow*)

Quartz crystals with their myriad uses for personal growth and
awareness are a fundamental part of 'the beings coming to life,
coming into their own, growing into their new expressive
selves'. They are helping us to grow in love and understanding
for our fellow beings, human, animal and plants, and our
planet, as we enter a New Age of heightened awareness. We
are rediscovering the knowledge of crystal power that our
ancestors once possessed long ago. Many people see this new
era as one that will bring peace, joy and love, a golden age in
which we will realize that, no matter what our colour, creed or
nationality is, we are all part of the same human family, inter-
related and deeply connected.

   Crystals are here to help us make the transition from one age
to the next. The Chinese call this time, '*Wei Chi*'. *Wei Chi*
means the moment of danger between cultures and the
opportunity to change that such a moment brings. It is the
death of the old and the birth of the new. At such a time we
experience the pains of death and the pangs of birth — the fear

of letting go of all that is known and familiar, combined with apprehension about what the future will bring.

Crystals can restore our balance, open our hearts and keep our spirits uplifted. In the latter part of the twentieth century we began to rediscover the practical uses of crystals – in computers, radios and televisions for example – but we still have a lot to learn about their secret powers and other ways in which they can improve our lives. By understanding how to use them we can heal the gap between each other and the Earth on which we live. We can build a bridge of communication to plants, animals and other dimensions.

Crystals understand our needs

One of the most magical attributes of gemstones is that they will often react to and mirror whatever is taking place in our lives. A young man in a workshop acquired two crystals and during the lunchbreak, took them to the beach to cleanse them in seawater. A wave washed one of them out of his hand and into the sea. He never found it again, was extremely upset and asked me why I thought it had happened. The crystal he lost was a male crystal and I said that, to me, it looked as if he needed to release his macho male energy and flow with his feelings. He had already decided to be more sensitive to his girlfriend. He grew up with a crystallized idea of what a man should be and realized that it was time to release it. The crystal mirrored this decision by floating into the sea, which symbolizes both spirit and feeling. When he understood this he felt better and released the crystal to its next task.

Another man, John, with whom I worked, was initially extremely tense and apprehensive. During the healing session he began to relax and suddenly one of my big crystals split in half. I was an amazed witness to the crystal's reflection of a major breakthrough. I gave him the piece that broke away as a reminder to him that he too could open up, break away and expand his life.

A friend of mind had a cluster of three crystals that were joined at the base. When her three children, almost simultaneously, left home the cluster broke into three separate stones. If this happens to your crystal it is important not to get upset. A broken crystal usually confirms what is going on in our lives and can sometimes give insight into what needs to happen.

Crystals and our Guardian Angels

Some crystals have horizontal lines across their length. These lines are called striations and you can both see and feel them as faint ridges. They represent steps to higher consciousness and can be very useful for developing your intuition. They also assist dreaming and meditation.

I have a number of crystals which, when you look into them, appear to have a steep and narrow path leading to a narrow door, shaped by the lines of the facets. These crystals also help us in our spiritual seeking. They will assist us to find, and stay on, the path of truth. If you have a project to complete and tend to get sidetracked this type of crystal can help you to be disciplined with your energy and say no to other distractions.

The crystal that is the most important to me, both personally and as a working crystal, is double-terminated, six inches long and full of rainbows. It has inside it the silvery shape of a small, kneeling figure with its hands raised in prayer. For me it is a reflection of a far greater power and looks exactly like an angel. I talk to this crystal angel constantly. I sometimes imagine or visualize myself inside this crystal. I ask questions and I always receive answers. These answers come in various ways. Sometimes words float into my mind or clear-cut scenes unfold in my head as if I were watching a movie. Often dreams will answer my question.

A friend of mine, Celia, used this method to ask which one of her family she had known before. (Celia believes in reincarnation.) The same night that she asked the question she dreamt that her little granddaughter, Hannah, was running towards her down a long tunnel, her arms outstretched and welcoming. At this time Hannah was only a baby and could not walk. Celia had always felt a special connection to this child and recognized the tunnel as being the same one she had herself traversed during a near death experience many years before. For Celia the dream was a direct answer to the question and her recognition of the tunnel confirmed to her that she had known her granddaughter in a previous life. Her belief in reincarnation came from everything that happened during her NDE.

If you want to make contact with your own guardian angel you can also imagine going inside, or connecting to, your crystal and then ask for that presence to join you. You do not have to have a figure already outlined in your crystal for this to work. Your subsequent communications will be of a telepathic nature. You can write your question on a piece of paper, or think it or even say it aloud. Afterwards write down any thoughts or feelings that come to you. If you make a practice of doing this you will gain much insight into your life.

Some years ago I presented a workshop in which one of the participants was an extremely arrogant young man whose entire conversation appeared to be criticism of the people in his life. He told me later that he decided to try my crystal method of communication because he did not believe it would work. He wanted to prove to me that it was far too simple. The day he tried it he wrote a letter to this — according to him, 'fantasy guiding angel' — and placed a crystal on top of the letter to help amplify and transmit the message. He asked for information about his life — was he doing the correct job, living in the best place, moving in the right circles? In a dream that night he saw himself staggering uphill carrying an enormously heavy load. His body was bent double and he felt such excruciating pain that he cried out, 'Why do I have to carry this burden, what am I being punished for?' A strong calm voice replied, 'This is the weight of the faults and the failures of the people around you. You noticed them, why should you not carry them?' His shock woke him up and he realised the justice of the reply. This resulted in such a dramatic change of personality that he came and spoke to me about it. He finally understood that people who are extremely critical of others are usually excessively judgemental of themselves first. The two are interconnected in the same way that we can only love others when we truly love ourselves first. He was then able to step out of his critical shell and become a very different, and likeable, man.

When I began communicating to and through the crystal

angel I believed that I must only ask questions of a spiritual nature. After a time, it occurred to me that my real friends were people on whose doors I'd knocked in a state of disarray when things had gone wrong. Usually I needed advice, money, help — even if only to borrow matches or milk. The friends who were more like acquaintances, the people I knew from high days, holidays, Christmas and birthdays, however nice, were not a familiar part of the basic, essential me. We met when we dressed in our best clothes, we showed each other the best side of ourselves. I then began communing with the angel constantly. I asked all sorts of mundane, everyday questions — even, 'where are my socks?' — 'what's happened to my left earring?' Through this kind of dialogue I developed a trust and rapport, just as we trust those people who see us and love us in our most scattered or bad-tempered moments.

In addition to conversations with my crystal-guardian-angel I also talk to the angel, or 'Spirit of Place', everywhere I go. I believe that each home, city and country has its own guiding angelic force and, that by communicating with this presence — greeting it when you arrive, thanking it when you leave — the place you live in, or visit, reveals itself to you in a different way.

You can do this by mentally projecting — or simply imagining — the words you want to say. If you hold a crystal in your right hand while doing this it amplifies the message. A result of this communication is that I sometimes feel led by invisible hands into unexpected spots of beauty or interest. It has also helped in a practical way. As a stranger in a city, I will find a restaurant under my nose when I need to eat, or the hotel that was fully booked suddenly has a space.

Through attunement to the Angel of New York I was able to walk alone through Central Park and Harlem. These are two places you are advised not to visit, especially on your own. In both cases I got off the subway too soon by mistake. It was funny, sad, extremely interesting and the only movement towards me was by a drunken woman who pinched my bottom. I smiled, she then smiled and, stepped back to allow me to pass. I felt totally protected by both the angel above me and the crystal in my pocket. The crystal amplifies our inner resources as well as the electro-magnetic field around us. It is like switching on a current that repels negativity.

171

This same process helped me on the London Underground when I was suddenly surrounded by a group of punk-rockers who were violently banging a young man's head against the platform. During the commotion a train arrived, I leapt inside, followed by the rockers, who closed me into my seat. Nasty little shivers of fright moved up my arms and legs but I clasped the crystal in my right pocket and began to talk, mentally — using my imagination — to the leader. I said, 'I do not know why you are behaving like this but you must have some reason. My soul can salute your soul, even if my personality is afraid.' I then directed thoughts of love and light towards him.

I had previously tried to get off the train but was not allowed to. Legs and bodies pushed me back into my seat. As I began to work with light and love from the crystal the leader's face relaxed. The train stopped again. I stood up, in preparation to get off, but their bodies, arms and legs still prevented me. I concentrated on the eyes of the leader, and he suddenly smiled, waved his henchmen away and said, 'It's OK, isn't it?' I smiled back, 'Yes it's OK,' and stepped from the train. The crystal activated my own energy; it also helped me to communicate with his inner essence rather than his outer personality. I am sure that both our angels, as well as the Angel of London, were there too.

I have a friend who hated her job, which involved a great deal of typing and mailing of letters. She began to use her crystal angel to imprint a message of love to everyone to whom a letter was sent. She suddenly saw herself as a healer, despatching healing energy through the letters she wrote. It changed her attitude to her job and also stimulated extra business for the company because the customers began to telephone in response to the letters. If we could all imprint love into every task we need to accomplish, life would move more smoothly, successfully.

Another friend used this attunement to achieve goals, such as taking her driving test. She asked her angel for protection and help and finally, after about eight previous attempts that failed, passed her test. Since then she has accomplished many other tasks by using the same formula of attunement to her angel.

The method can also help to heal various countries in the

world. Place a crystal on a map of the world and generate healing to the world or to the country most in need at this time. I have done this with groups of friends where we also arranged to check the results. We took a city or village for a week at a time. We monitored local emergency and social services, such as police, hospital, doctor and fire-station. In each case there was a significant drop in fire, crime and accident rates.

When I read a book or listen to music I use my crystal to attune through my angel to both the writer and the spirit of the composition. This gives me a deeper understanding of what the words and the music are saying. The crystal amplifies the interpretation.

If you want to start a conversation with your own inner teacher or guardian angel, using a crystal as a medium through which you converse is a good way to begin. You can of course do this without a crystal but crystal energy activates and facilitates the process. Always remember to thank and bless the crystal, as well as your angel, when you've finished.

Crystals and our future

As we move towards the end of the century a polarization is taking place of good and evil, both in the individual and universally. It is a little like making soup stock. You simmer the bones and a froth floats to the surface. Before the heat was applied the water looked pure. Now, when you scoop away the froth, you will find clear liquid underneath. In the same way our petty little negativities are being brought to the surface of our awareness to be cleared away. As long as they remain hidden we cannot do anything about them. Gems and crystals are here to implement the healing that leads to self-love and acceptance. Whether you deliberately programme your crystal or simply carry it around with you it will still have this effect.

At this time of waking up we are being pushed to examine every facet of life that we may previously have taken for granted. People who feel neglected can become ill to call for love and help. Plants when hated, loved or ignored die first when ignored. The Earth has, for too long, been ignored and now manifests symptoms of dis-ease as it screams for our attention, through earthquakes, flood and famine.

Crystals are an integral part of the Earth's core. They are here to help us listen, not only to each other but to what the planetary being that is the Earth most needs at this time. Crystals placed in or on the earth will facilitate this. You can also attune to the Earth, and send healing to it through the crystal you hold in your hand. I have found rose quartz particularly good for Earth healing; it is soft, gentle and especially beneficial for areas that have been raped through man's insensitivity (for example, building roads through rain forests; plundering the Earth for fuel, taking without putting back).

The heart chakra of the Earth is opening — 'It is the time of manifesting the dreams that were brought to us during the time, of the North, the Winter. As you look around you will see that it is a time of love. A time to watch all creatures doing their dance of life.' We must listen with our hearts rather than with our own heads, understand with our feelings and intuition instead of our logic.

Part of this flow of feeling is enabling men and women to develop a different relationship with each other, a relationship of empathy and companionship rather than that of dominator and slave. We were all originally androgynous, a combination of male and female, neither one predominant. We are moving back to androgyny which is why there is so much sexual confusion at this time, men who play more passive female roles, women asserting themselves like men.

In the legend of Psyche and Aphrodite Psyche was born when a dewdrop from Heaven fell to Earth. Therefore Psyche symbolizes the fragile, vulnerable side of a woman. Aphrodite was born when the seeds of Uranus were scattered on the oceans of the world, she came out of the depths and symbolizes the deep primeval power of the wise, nurturing woman, the priestess who can also become the witch-bitch if scorned or ignored.

All over the world today I see a warring between the Psyche and Aphrodite in a woman, as well as in the inner woman in a man. Aphrodite in each of us, male and female, is clamouring to be heard. She tugs and shoves at our entrails and says, 'get up — show yourself — do something'. The woman feels impelled from within to stand up and flex her muscles a little. Suddenly her inner Psyche pops up and says, with a trembling voice, 'I don't know what to do and even if I did I could not do it. I'm afraid, leave me alone,' and the woman sinks back, feeling frustrated and helpless to help herself. In a man Aphrodite pushes him to respond from feeling, emotion rather than with logic. Often when he does he frightens himself, he may lose control, the Psyche in him says, 'I told you so. Your feelings will run away. It's frightening, it doesn't work, don't do it again.'

As the Psyche-Aphrodite battle works itself out in men and women — when we stop using each other as supportive props — we can become whole within ourselves, sometimes assertive and sometimes receptive. Crystals can help bring us to this point of balance so that we can talk to each other as androgynous, fellow, human-beings instead of trying to fulfil pre-ordained male/female roles.

The Earth is undergoing an initiation and we who are on the Earth are experiencing different levels of that same initiation. It is an initiation of love, and crystals — being a manifestation of love and light — are here to play a vital role. They are here to wake us up, make us more sensitive, remind us of our true identity and where that identity comes from. I predict that in the future, like the Atlanteans before us, we will use crystal energy in the same way that we use electricity today.

We are only just beginning to rediscover the power of the mineral kingdom. I believe knowledge is contained within stones such as those at Stonehenge, Avebury, Machu Pichu in Peru, the Mayan temples in Guatemala, the pyramids of Egypt and Mexico, the stone temples in Angor Wat, Cambodia and many other well-known sites. I feel this knowledge was imprinted by wise men, just as we imprint information on to micro-discs today. Through science as well as religion we are beginning to read these stones. Crystals provide a bridge between right and left brain, yin and yang, positive and negative, male and female, science and spirit, and will in the future, I believe, enable them to merge.

In Los Angeles I have a friend who owns a stone that she somewhat reluctantly purchased from a store, prompted by a nudge from the stone to 'buy me, buy me'. This stone is of a fairly insignificant variety and small enough to fit into the palm of your hand. She took it home and placed it on her mantel-

Crystal Cluster

piece, thought no more about it until a visiting psychic told her that it had been at the foot of the Cross and contained within it some drops of Jesus's blood. During the next six months a number of psychics, independently and without being told what the previous clairvoyant had said, all told the same story.

Suddenly the stone vanished. After some weeks my friend, with another girl, was sitting in her garden when she saw an object flying through the air towards them. It hovered within six feet of where they sat and they suddenly saw it was the stone. Too paralysed to move they watched it disappear around the house before they leapt to their feet and followed it. It was nowhere to be seen but later that evening they found it back on its place on the mantelpiece. They immediately contacted the original psychic and asked where the stone had been. Over the next few weeks they asked many other psychics. Each one again told the identical story. This was that the stone had been to Venus, where Jesus is supposed to have come from. It would now remain on Earth until the consciousness of love and the Holy Spirit — some know it as the Christ consciousness — was truly anchored into the Earth and the hearts and minds of all mankind. When this happens the stone will shatter, its work completed.

Stones, gems and crystals are magical and mysterious, a law unto themselves. They appear and disappear, lose their lustre when we are ill and sparkle when we are happy. They can be used to amplify thought, impart information, focus and transmit energy, generate love. Crystals are a rainbow bridge by which we can freely and joyously participate in the dance of life.

Walking Buffalo said, 'Did you know the trees talk? Well, they do. They talk to each other, and they'll talk to you if you listen. The trouble is white people don't listen. They never learned to listen to the Indians, so I don't suppose they'll listen to the other voices in nature. But I have learned a lot from trees.'

Crystals talk. They'll talk to you if you listen. As you walk upon the Earth they will help you listen to all the beings coming to life, all the voices in nature. I have learnt a lot from crystals, especially about love — I hope you will too.

Index

Useful Addresses of Crystal Suppliers and Shops

London

Crystals
9 Adelaide St
Strand
London WC2

The Mystic Trader
60 Chalk Farm Road
London
NW1 8AN
Tel: 01-284-0141
Do mail order

Aurora Crystals
16a Neal's Yard
London WC2H 9DP
Tel: 01-379-0818
Mail order for lead crystals only

Gloucestershire

Reflections of Rainbows
John Street
Stroud
Gloucestershire GL5 2HA
Tel: 045-367-2929
Do mail order

Lancashire

The Crystal Research
    Foundation
37 Bromley Road
St Annes on Sea
Lancashire FY8 1PQ
Tel: 0253-723735

Leicestershire

Crystal World
Anubis House
9 Creswell Drive
Ravenstone
Leicestershire LE6 2AG
Tel: 0530-510864

Scotland

Orcadian Stone Company Ltd
Golspie
Sutherland
KW10 6RH
Tel: 040-83-3483
Do mail order

Somerset

Opie Gems
57 East Street
Ilminster
Somerset
Tel: 0460-52346
Do mail order

Sussex

The Unicorn Place
39 Duke Street
Brighton
Sussex BN1 1HE
Tel: 0460-52346
Do mail order

South Africa

The Wellstead
1 Wellington Avenue
Wynberg
Cape 7300
Tel: 797-8982
Do mail order

Topstone Mining
  Corporation C.C.
Dido Valley Road
PO Box 20
Simonstown 7975
Tel: (021) 86-2020/1/2/3
Do mail order

Rock and Gem Shop
London Arcade
Durban
Tel: 304-9267

Australia

The Mystic Trader
125 Flinders Lane
Melbourne 3000
Australia
Tel: 03-650-4477
Do mail order

The Rock Shop
Arcade 83
Shop 4
83 Longueville Road
Lane Cove
Sydney NSW 2066
Tel: 428-4247

New Zealand

Moa Unlimited
413 Richmond Road
Greylynn
Tel: 09765-065

Gem Rock and Minerals
52 Upper Queen Street
Auckland
Tel: 09-774-974

Rotorua Lapidary Rock and
    Mineral Supplies Shop
29 Geyser Court
Rotorua
Tel: 073-88996

Wilderness Gems Ltd
13 River Road
Ngatea
Tel: 0843-77417

Further Reading

Books on Crystals

Cosmic Crystals Ra Bonewitz, Element Books (1986)

The Mystical Lore of Precious Stones Vols I & II George Frederick Kunz, Newcastle Pub. Co., (1986)

The Diamond George Blakey, Paddington Press (1977)

Books on Spiritual Development

Illusions Richard Bach, Pan Books (1979)

Jonathan Livingstone Seagull Richard Bach, Pan Books (1973)

Bridge Across Forever Richard Bach, Pan Books (1985)

World Within Gina Cerminara, C.W. Daniel (1973)

The Aquarian Conspiracy Marilyn Ferguson, Paladin (1982)

Lives and Teachings of Masters of the East B.T. Spalding, De Vorss, U.S.

Hidden Wisdom in the Holy Bible Geoffrey Hodson, Theosophical Publishing House (1974)

The Tao of Pooh Hoff, Magnet Books (1984)

Mr God, This is Anna "Fynn", Found Pub. (1977)

Edgar Cayce Speaks Edgar Cayce, Avon Bks., U.S. (1987)

Notebooks Paul Brunton, Larson Bks. (1977)

Silent Paths Michel Eastcott, Rider (1987)

Behold the Spirit Alan Watts, Vintage Bks., U.S. (1973)

Essential Alan Watts Alan Watts, Celestial Arts (1985)

Meaning of Happiness Alan Watts, Rider (1978)

Nature, Man and Woman Alan Watts, Vintage Bks., U.S. (1988)

Supreme Identity Alan Watts, Vintage Bks., U.S. (1988)

This is It Alan Watts, Rider (1978)

Testimony of Light Helen Greaves, Spearman (1969)

Initiation Elizabeth Haich, Unwin (1974)

Life After Life Raymond Moody, Bantam (1983)

Autobiography of a Yogi Paramahansa Yogananda, Rider (1987)

Love is Letting Go of Fear Jerry Jampolski, Celestial Arts (1982)

Out on a Limb Shirley MacLaine, Bantam (1987)

Siddartha Herman Hesse, Picador (1974)

Way of the Sufis Idries Shah, Penguin (1974)